Introduction
to
Statistics

INTERNATIONAL SERIES IN DECISION PROCESSES

INGRAM OLKIN, Consulting Editor

A Basic Course in Statistics, 2d ed., T. R. Anderson and M. Zelditch, Jr.
Introduction to Statistics, R. A. Hultquist
Applied Probability, W. A. Thompson, Jr.
Elementary Statistical Methods, 3d ed., H. M. Walker and J. Lev
Reliability Handbook, B. A. Koslov and I. A. Ushakov (edited by J. T. Rosenblatt and L. H. Koopmans)
Fundamental Research Statistics for the Behavioral Sciences, J. T. Roscoe

FORTHCOMING TITLES

Introductory Probability, C. Derman, L. Gleser, and I. Olkin
Probability Theory, Y. S. Chow and H. Teicher
Statistics for Business Administration, W. L. Hays and R. L. Winkler
Statistical Inference, 2d ed., H. M. Walker and J. Lev
Statistics for Psychologists, 2d ed., W. L. Hays
Decision Theory for Business, D. Feldman and E. Seiden
Analysis and Design of Experiments, M. Zelen
Time Series Analysis, D. Brillinger
Statistics Handbook, C. Derman, L. Gleser, G. H. Golub, G. J. Lieberman, I. Olkin, A. Madansky, and M. Sobel

Introduction to Statistics

Robert A. Hultquist
Pennsylvania State University

HOLT, RINEHART AND WINSTON, INC.

New York · Chicago · San Francisco · Atlanta · Dallas
Montreal · Toronto · London · Sydney

**TO MY WIFE, LYNNE
AND MY MOTHER, ALICE**

*When I applied my heart to know wisdom,
and to see the business that is done upon the earth
then I beheld all the work of God,
that man cannot find out the work that is done
under the sun: because however much a man labor
to seek it out, yet he shall not find it;
yea moreover, though a wise man think to know it,
yet shall he not be able to find it.*

ECCLESIASTES 8: 16, 17

1492

PREFACE

This text is the outgrowth of a careful editing of lecture notes, laboratory exercises and quiz problems given, during the past ten years, to students in first courses in statistical methods. It is a text which perhaps reflects more of what faculty members, outside of mathematical science, feel is important than what I personally like to teach. It is a text for students with motivation to perform experiments and to analyze the results, but I have found that it is also a text for beginning students of mathematical statistics. A professional statistician without a good knowledge of the concepts presented in this text would be considered by many of his colleagues to have a serious gap in his training.

Great effort has been put forth to make the text one unified story. The restriction to consider only sampling from normally distributed populations was made in order not to interrupt the plot. This restriction helped make it possible to present many concepts in a relatively small volume. I believe that in a first course it is better to examine many concepts with reference to one parent population than to examine a few concepts with respect to many parent populations. It is hoped that through thoroughly understanding the role of the normal density the student will, by transfer of knowledge, be in a position to efficiently learn statistical procedures for situations associated with other density functions.

The text is devised so that classical high school algebra is the only mathematical prerequisite. Early in the text the reader is introduced to summation notation, and this notation is used frequently throughout the remainder of the text. The topics are those which can be easily handled without matrix notation although the use of matrix notation would in many instances make the presentation more elegant. Although the topics are for the most part those taught in most beginning methods courses, the emphasis given to many of the topics is different from that given to these topics in most courses. For example, the formal presentation of probability is confined to a few pages.

The text starts with a discussion of some very basic concepts which lead into an introduction of inference. The story of statistical inference proceeds rather rapidly until it reaches analysis of variance and regression, whereupon topics are more carefully surveyed. This approach is preferred by most experimental and consulting statisticians. After the basic concepts are presented, the text deals with point estimation methods, interval estimation methods, and tests of statistical hypotheses in association with many different experimental situations. Some emphasis is placed on choosing a model to adequately represent the data, and in this regard the text employs a rather different illustration technique. For several models, observations are created from assumed parametric values and then the created observations are analyzed. Finally, the assumed values of the parameters are compared with the estimate obtained from created data.

The text can be used in a course with or without a weekly laboratory session. Numerous problems are placed in each chapter, and some of the problems are designated "Laboratory Type Problems." Most of the problems are an integral part of the text, and it is strongly recommended that all or nearly all of them be assigned as homework or worked in class. Since expected mean squares play a very important role in inference theory, numerous problems associated with determining expected mean squares have been included in the exercise sections. Students generally find these problems exceedingly difficult; hence, it is advised that some of these problems be worked by the instructor. If most of the problems are worked, the text contains material for a two-term course.

The text is not specifically designed for any discipline-oriented class. It contains illustrative examples and exercises from a variety of application areas. Students usually appreciate a statistical method to a greater extent when they see the power associated with wide applicability of a procedure. Few if any of the illustrations are so technically discipline-oriented that a student outside that discipline gets lost in the jargon of the illustration.

I would like to thank each senior member of the Statistical Laboratory at Oklahoma State University for educating me in experimental design and instilling in me a desire to help experimenters. Thanks go to the faculty and staff of Pennsylvania State University and to the chairman of that department for providing an academic climate in which I had the time and desire to prepare the manuscript. I appreciate the efforts of about a dozen individuals at Pennsylvania State University and Holt, Rinehart and Winston, Inc. who typed and in other ways worked on the manuscript. Thanks go to my immediate family and to my in-laws, who encouraged me to write this book. Finally, I acknowledge with deep appreciation my parents, who among many things made my professional education possible.

Boalsburg, Pennsylvania   Robert A. Hultquist
March 1969

CONTENTS

Preface   vii

1 Basic Concepts   1

   1.1   Introduction, 1
   1.2   Typical Values, 2
   1.3   Some Properties of Summation Notation, 4
   1.4   Exercises, 5
   1.5   Measures of the Variability in Data, 6
   1.6   Exercises, 8
   1.7   Random Variables, Histograms, and Distributions, 9
   1.8   An Elementary Discussion of Probability and Its Properties, 11
   1.9   Exercises Involving the Standardized Normal Table, 13
   1.10 The Normal Curve, 13
   1.11 Exercises Involving Normal Variables, 16
   1.12 Degrees of Freedom, 16

2 Statistical Inference Relative to Sampling from One
   Normal Population   19

   2.1   Introduction to Sampling, 19
   2.2   The Nature of $\bar{X}$, 20
   2.3   An Example to Illustrate the Variability of $\bar{X}$, 22
   2.4   Computational Exercises Related to the Distribution of $\bar{X}$, 23
   2.5   A Sample-size Problem, 24
   2.6   Exercises in Determining Approximate Sample Sizes, 25
   2.7   The Concept of an Interval Estimate, 25
   2.8   Exercises Involving Interval Estimates of μ When σ^2 Is Known, 27
   2.9   The Concept of Testing a Statistical Hypothesis, 28

2.10   Exercises Involving the Test of a Simple Hypothesis about μ When σ^2 Is Known, 32

2.11   Student's t Distribution and Its Role in Statistical Inference, 33

2.12   Exercises Involving Student's Table, 35

2.13   Composite Hypotheses, 36

2.14   Exercises, 39

2.15   The Chi Square Distribution and Some Applications, 40

2.16   Exercises Involving Inference Relative to the Variance, 45

2.17   Point Estimation, 46

2.18   Exercises, 51

3   Statistical Inference Relative to Sampling from Two Normal Populations   54

3.1   Introduction to the Study of Two Populations, 54

3.2   Weighing Estimates of μ and σ^2, 56

3.3   Exercises, 57

3.4   Inference Relative to the Means When Independent Samples Are Taken and Equality of Variance Is Assumed, 60

3.5   Exercises, 62

3.6   The Snedecor F Distribution and Some Applications, 63

3.7   Exercises, 65

3.8   Paired Experiments and the Concept of Experimental Design, 67

3.9   Exercises, 69

4   Statistical Inference Relative to Sampling from More Than Two Normal Populations   71

4.1   Introduction to the Study of More Than Two Populations, 71

4.2   The Completely Randomized Experimental Design, 73

4.3   Exercises, 75

4.4   The Concept of a Linear Statistical Model, 77

4.5   One-way Classification Experiments with Unequal Numbers Per Classification, 80

4.6   Exercises, 81

4.7   Two-way Cross-classification Designs, 84

4.8   Exercises Related to the Two-way Cross-classification Type of Experiment, 87

5   Basic Concepts When Two Characteristics Are Studied   90

5.1   Introduction, 90

5.2   The Concept of Covariance and the Concept of Linear Correlation, 91

5.3   Exercises, 92

5.4   Introduction to Least Squares, 92
5.5   Exercises, 96
5.6   Mathematical Models for Regression, 97
5.7   The Regression AOV and Tests of Hypotheses, 101
5.8   An Academic Illustration, 105
5.9   Exercises, 107
5.10   The Case Where a Linear Regression Model May Not Fit, 109
5.11   Adjusted Scores, 110
5.12   Exercises, 111

6   Statistical Inference When Interaction Is Present   113

6.1   The Concept of Interaction, 113
6.2   Factors and Contrasts, 114
6.3   Statistical Models When Interaction Is Present, 115
6.4   Testing Hypotheses and the AOV for 2^2 Factorial
        Experiments, 117
6.5   Exercises, 120
6.6 2^3 Factorial Experiments, 122
6.7   Exercises, 125
6.8 3^2 Factorial Experiments, 125
6.9   The Role of Factorial Experimentation, 132
6.10   Exercises, 134

7   Statistical Inference for Some Special Models   138

7.1   Introduction, 138
7.2   The Two-fold Hierarchal Model, 141
7.3   An Illustrative Two-fold Hierarchal Problem, 144
7.4   Exercises, 145
7.5 N-way Cross-classification Designs, 147
7.6   Exercises, 150
7.7   Analysis of Incomplete Data, 153
7.8   Exercises, 157

Appendix   159

Table I   The Standard Normal Distribution, 161
      II   The Chi Square Distribution, 164
      III   The Student t Distribution, 165
      IV   The Snedecor F Distribution, 166
      V   Gaussian Deviates (Values of Z), 174
      VI   Gaussian Deviates (Values of X), 179

Bibliography and References   185

Index   187

Bibliography and References, 185
Index, 187

1

BASIC CONCEPTS

1.1 INTRODUCTION

Statistics is a science that concerns itself with experimentation and the collection, description, and analysis of data. The *statistical layout* displayed in Table 1.1 contains data typical of that considered by many

Table 1.1 Strength Test Scores of Sophomore Boys by High School and Class

High School	1	2	3	4	5
Scores from class $(1, j)$	70.3	78.9	52.3	69.8	80.5
	80.9	83.1	59.8	58.5	88.3
	59.7	85.2	61.3	75.7	81.2
	63.8	64.7	70.6	59.6	79.5
	58.5		71.7	68.7	74.4
	75.3		75.4	81.3	
Average for class $(1, j)$	68.1	78.0	65.2	68.9	80.8
Scores from class $(2, j)$	60.5	73.8	72.3	59.8	*No second class*
	61.3	78.5	77.6	65.4	*in this school*
	62.8	83.7	78.9	83.1	
	69.7	84.5	69.0	57.6	
	58.3	78.9	83.2	55.2	
	65.5	77.5	69.0	48.8	
	72.8	79.6		57.3	
Average for class $(2, j)$	65.2	79.5	75.0	64.5	

1

statisticians. The test scores recorded in Table 1.1 correspond to strength tests given to boys in nine different sophomore physical education classes. The boys and the scores that they made are examples of what more generally are called *experimental units* and *observations*, respectively. Five different schools were represented with two classes from each of four schools and different teachers for each class. The teacher (class) designation is as follows: (1, 3) indicates the first teacher from the third high school; (2, 4) denotes the second teacher from the fourth high school; and, in general, (i, j) denotes the ith teacher from the jth high school.

With reference to the data, we might ask: Do the scores provide evidence that one school is better than the rest with respect to its physical education program? Are there significant differences in teachers? Do some classes have significantly more variability than other classes? What is a typical score for the class with teacher (i, j)? There are of course many other questions that might be asked. Some of the questions are even more basic, for they deal with concepts at the very heart of experimentation. Are there enough boys to give good answers to questions such as those above? Is there a better layout for displaying the results of the experiment? What type of new experiment should be run if more information is desired relative to the teachers, the high schools, and the physical fitness of the students? This book presents methods that attempt to answer questions such as those posed here. It presents statistical methods for describing and analyzing results of experiments, augmented by discussions of why a technique is a good one and why it is preferred over another technique in a given situation.

Statistical methods are tools for examining data. The statistician dealing with business trends, the statistician keeping the baseball average up to date, the statistician conducting public opinion polls, these and many others, all possess a common characteristic in that they deal with data. Data handlers do not, of course, use the same tools all of the time, but statistical methods preferred by different types of statisticians do overlap, and it is the overlapping concepts that are stressed in this treatment of statistics. With these comments in mind, we shall consider some of the basic concepts underlying the study of data.

1.2 TYPICAL VALUES

Data can be and are presented in several ways. We consider here a numerical description of the outcome of an experiment. The word "experiment" is used to include the case where we merely observe nature. Let the numbers (observations), say n of them, that come from an experiment be denoted by indexed letter symbols such as X_1, X_2, $\ldots$, X_n or Y_1, Y_2, $\ldots$, Y_n. In the study of numerical data, perhaps the most elementary and, in many respects, the most important concept is that of the *arithmetic mean* or average of n observations. Convention dictates that the *arithmetic mean*, henceforth referred to as the mean, be denoted by the symbol in use, covered by a bar, and the *mean* of Y_1, Y_2, $\ldots$, Y_n is defined to be $\bar{Y} = (Y_1 + Y_2 + Y_3 + \cdots + Y_n)/n$.

It is convenient to denote a sum such as $Y_1 + Y_2 + \cdots + Y_n$ by the symbols $\sum\limits_{i=1}^{n} Y_i$ which are read, "the sum of Y_i from i equals 1 to i equals n." The Greek letter Σ (sigma) is referred to as the *summation sign*, and in summation notation the mean of Y_1, Y_2, $\ldots$, Y_n becomes

$$\bar{Y} = \frac{1}{n} \sum_{i=1}^{n} Y_i.$$

$\bar{Y}$ is a measure of the *typical value* for the observations Y_1, Y_2, $\ldots$, Y_n, and although there are other measures of the typical value, we shall confine our attention to the mean.

The term *population* is used in statistical literature to denote the set of all possible experimental units or the set of all numerical values corresponding to the experimental units. A subset of a population of experimental units, or the observations Y_1, Y_2, $\ldots$, Y_n corresponding to a characteristic that they possess, is called a *sample* from the population. The number n of experimental units in the sample is called the sample size. To illustrate the idea, consider the three distances, recorded in feet, for a shot putter in a track and field event: $Y_1 = 57.6$, $Y_2 = 52.4$, and $Y_3 = 59.2$. Conceivably, there are a great number of different distances that might have been attained. The three sample values recorded constitute a subset of the set of all possible distances that might have been attained. $\bar{Y}$ in this illustration is 56.4. Behind the scenes there is another number playing the role of the typical value for the population of all possible distances. This number we denote by the Greek letter μ and it is called the *population mean* or the *expected value* of the population.

In statistical studies, the word *parameter* is used to convey the idea of a fixed quantity in a given experimental situation, with the understanding that if the experimental situation is changed, the quantity may take on another fixed value. The population mean μ is an example of a parameter. If the experimental situation is changed, for example, by changing the weight of the shot in the track and field meet, then the mean of the population of all distances may take on another fixed value.

1.3 SOME PROPERTIES OF SUMMATION NOTATION

As indicated earlier, $\sum_{i=1}^{n} Y_i$ denotes $Y_1 + Y_2 + \cdots + Y_n$. Contemplating the state of affairs, a mathematician one day commented that in a sense, much of mathematics appears to be a science of position. What he had in mind can be illustrated by referring to the symbols $\frac{X}{2}$, X_2, X^2, $2X$, and $2/X$. The relative positions of the 2 and the X are indeed very important to the understanding of what the writer has to say. It might also be observed that we really are not dealing with concepts here but with definitions and language.

In discussing the symbols $\sum_{i=1}^{n} Y_i$, we refer to i as the index of summation and say that the index runs from 1 to n. Some other shorthand notations that are really extensions or different ways of looking at summation notation are the following:

$$\sum_{i=1}^{k} X^i = X + X^2 + X^3 + \cdots + X^k$$

$$\sum_{i=a}^{b} i = a + (a + 1) + (a + 2) + \cdots + (b - 1) + b$$

$$\sum_{i=1}^{n} (X_i + Z_i) = (X_1 + Z_1) + (X_2 + Z_2) + \cdots + (X_n + Z_n)$$

$$\sum_{i=k}^{m} cX_i = cX_k + cX_{k+1} + \cdots + cX_m.$$

After the meaning of summation notation is understood and before we try to use the notation efficiently, it is important to learn and then to recognize basic arithmetic properties in this notation. For example,

$$\sum_{i=1}^{n} (X_i + Z_i) = \sum_{i=1}^{n} X_i + \sum_{i=1}^{n} Z_i.$$

This property will now be verified after which several similar properties, listed in the following exercises, should be verified by using basic arithmetic properties of numbers.

By definition,

$$\sum_{i=1}^{n} (X_i + Z_i) = (X_1 + Z_1) + (X_2 + Z_2) + \cdots + (X_n + Z_n).$$

By the associative law of arithmetic,

$$(X_1 + Z_1) + \cdots + (X_n + Z_n) = X_1 + Z_1 + X_2 + Z_2 + \cdots$$
$$+ X_n + Z_n = (X_1 + X_2 + \cdots + X_n) + (Z_1 + \cdots + Z_n).$$

But

$$\sum_{i=1}^{n} X_i = (X_1 + X_2 + \cdots + X_n)$$

and

$$\sum_{i=1}^{n} Z_i = (Z_1 + Z_2 + \cdots + Z_n).$$

Thus, by substitution, we have

$$\sum_{i=1}^{n} (X_i + Z_i) = \sum_{i=1}^{n} X_i + \sum_{i=1}^{n} Z_i.$$

1.4 EXERCISES

Verify that

(1) $$\sum_{i=1}^{n} (X_i - Z_i) = \sum_{i=1}^{n} X_i - \sum_{i=1}^{n} Z_i$$

(2) $$\sum_{i=1}^{n} cX_i = c \sum_{i=1}^{n} X_i$$

(3) $$\sum_{i=1}^{n} c(X_i + Y_i + Z_i) = c \sum_{i=1}^{n} X_i + c \sum_{i=1}^{n} Y_i + c \sum_{i=1}^{n} Z_i$$

(4) $$\sum_{i=1}^{n} c = nc$$

(5) $$\sum_{i=1}^{n} i = \frac{n(n+1)}{2}$$

(6) $\displaystyle\sum_{i=1}^{n} X_i = n\bar{X}$

(7) $\displaystyle\sum_{i=1}^{n} (X_i - \bar{X}) = 0$

1.5 MEASURES OF THE VARIABILITY IN DATA

Unless each observation Y_i in a set of observations is numerically equal to each and every other observation, there is variability in the sample. In a very real sense, the statistician's business is to study variability. There are many measures of the variability of a sample, including the easily computed *range* and the *average absolute deviation* from the mean. By definition, range = {maximum Y_i} − {minimum Y_i} and average absolute deviation from the mean equals

$$\frac{1}{n} \sum_{i=1}^{n} |Y_i - \bar{Y}|$$

where $|Y_i - \bar{Y}| = Y_i - \bar{Y}$ if Y_i is greater than $\bar{Y}$ and $|Y_i - \bar{Y}| = -(Y_i - \bar{Y})$ if Y_i is less than $\bar{Y}$. Notice that, since $\displaystyle\sum_{i=1}^{n} (Y_i - \bar{Y})$ always equals zero, the average deviation from $\bar{Y}$ is not a candidate for measuring variability.

For most experimental situations, the *variance* is the measure of variability with properties that far surpass the properties of other measures of variability; hence, the variance or its positive square root is the measure most often used. The *sample variance* associated with the observations $Y_1, Y_2, \ldots, Y_n$ will be denoted by S^2 or S^2_Y and is defined by the equation

$$S^2_Y = \frac{1}{n-1} \sum_{i=1}^{n} (Y_i - \bar{Y})^2.$$

There are good reasons for dividing by $(n-1)$ instead of using the average squared deviation, but these reasons will be presented later because they involve a concept not yet introduced.

The population counterpart of S^2_Y (that is, the measure of variability in the population from which the sample was taken) is called the population variance. It is a parameter that we shall denote by σ^2_Y or σ^2. The square roots of S^2 and σ^2 are called the sample and population *standard deviations,* respectively.

The notion of variance will first be studied by examining some, perhaps artificial, data that is convenient to work with. This is done in order to familiarize the reader with important properties of the variance and standard deviation measures of variability. Consider the following sample of observations made, say, in yards:

$$Y_1 = 2, \quad Y_2 = 5, \quad Y_3 = 6, \quad Y_4 = 5, \quad Y_5 = 3.$$

For this data, $\bar{Y}$ is 4.2 and $S^2_Y = \frac{1}{4} \sum_{i=1}^{5} (Y_i - 4.2)^2$ works out to be 2.7.

Now consider the same data recorded in feet instead of yards. If $X_i = 3Y_i$, then $X_1 = 6$, $X_2 = 15$, $X_3 = 18$, $X_4 = 15$, and $X_5 = 9$. $\bar{X}$ for the data, in terms of feet, is 12.6 and S^2_X, as the student can verify, becomes 24.3. It may be observed that $\bar{X} = 3\bar{Y}$ and $S^2_X = 3^2 S^2_Y$. This is not an arithmetic accident. The general statement in theorem form is:

Theorem 1.1 *If for a sample of size n we have $X_i = cY_i, i = 1, 2, \ldots, n$, then $\bar{X} = c\bar{Y}$ and $S^2_X = c^2 S^2_Y$.* The problems in the next section guide the student through a proof of this very important theorem.

The statement of this theorem implies that a sample variance is of little value in itself and, indeed, we shall find later that a sample variance is almost always used in conjunction with another sample variance and/or a sample mean. Since the scale of measurement influences the numerical value of a variance for a sample, a variance should be considered large or small only in respect to other functions of observations measured in the same scale.

In applying the formula

$$S^2_Y = \sum_{i=1}^{n} \frac{(Y_i - \bar{Y})^2}{n - 1}$$

it is necessary to subtract n different times. Even in the case where the Y_i are convenient one- or two-digit numbers, the quantity to be squared may not be nearly so convenient to work with. Especially when n is large, equivalent expressions for S^2_Y may be easier to apply. Two equivalent expressions are

$$S^2_Y = \frac{1}{n - 1} \left[\sum_{i=1}^{n} Y_i^2 - n\bar{Y}^2 \right]$$

and

$$S^2_Y = \frac{1}{n-1}\left[\sum_{i=1}^{n} Y_i^2 - \frac{1}{n}\left(\sum_{i=1}^{n} Y_i\right)^2\right].$$

They are often referred to as computational forms of the sample variance. Observe that these forms demand the subtraction operation only once. If calculations are done with the aid of a hand computer, the observations Y_i need be entered only once, for ΣY^2_i and ΣY_i may be obtained simultaneously.

1.6 EXERCISES

(1) For the following data, compute the mean, the range, the average absolute deviation, the sample variance by using the definition of S^2_Y, and the sample variance by using a computational form of S^2_Y:

$$Y_1 = 7, \qquad Y_2 = 3, \qquad Y_3 = 4, \qquad Y_4 = 5, \qquad Y_5 = 3, \qquad Y_6 = 4.$$

(2) Let $X_i = 2Y_i$ for each Y_i in Exercise (1), then verify that $S^2_X = 4S^2_Y$ by evaluating S^2_X according to the definition and by applying a computational form for S^2_X.

(3) Show that by applying the proper techniques, the computations for finding S^2_Y for the data $Y_1 = \frac{1}{3}$, $Y_2 = \frac{4}{3}$, $Y_3 = \frac{2}{3}$, and $Y_4 = 1$, can be made quite easy.

(4) When $X_i = cY_i$ for $i = 1, 2, \ldots, n$, prove that $\bar{X} = c\bar{Y}$ and $S^2_X = c^2S^2_Y$ by writing the definitions of $\bar{X}$ and S^2_X and making the substitution $X_i = cY_i$.

(5) Working from the definition, prove that

$$S^2_Y = \frac{1}{n-1}\left[\sum_{i=1}^{n} Y_i^2 - n\bar{Y}^2\right].$$

(6) Prove that $\sum_{i=1}^{n} (X_i - a)^2 \geq \sum_{i=1}^{n} (X_i - \bar{X})^2$ for every value of a.

(7) A laboratory-type problem: Each of several students should select at random six observations (numbers) from Table V. For each sample of six observations, compute S^2_Y and tabulate the results.

(8) A laboratory-type problem: Each of several students should select at random 18 observations from Table V. Compute and tabulate the values of S^2_Y obtained from these samples. Compare the results with those obtained when sample size six was used and with the value of σ^2 reported in the table.

1.7   RANDOM VARIABLES, HISTOGRAMS, AND DISTRIBUTIONS

When a track and field participant is about to put a shot, the distance the shot will travel is unknown and is referred to by statisticians as a *random variable*. After the event, the recorded distance can be thought of as an observational value taken on by a *random variable*. The birth weight of an unborn child is another example of a random variable. The recorded birth weight for a particular child can be thought of as an observation taken on by the birth-weight random variable. The important relationships between samples and random variables will be investigated by considering an example. Whereas, for the most part throughout this book, populations with great numbers of experimental units are considered, in this example the population consists of five experimental units labeled A, B, C, D, and E. If two different experimental units are selected by some chance procedure, then ten different samples can result. They are AB, AC, AD, AE, BC, BD, BE, CD, CE, and DE. If the chance procedure is such that each pair has an equal chance of being selected (for example, drawing two of the five letters from a hat), then the pair that occurs is said to be a *random sample*[1] of size two. When k experimental units are selected from a population of p experimental units in such a way that each set of k units has an equal chance of being selected, the particular set of units obtained is called a random sample of size k. When a random sample of size one is selected, the numerical value Y, for some characteristic of the experimental unit, is another example of a random variable.

A concrete illustration of a random variable of the type just mentioned is the weight of a randomly selected leaf on a particular tree. This illustration will be developed in some detail. Consider a random sample of six leaves from a tree. We wish to diagram the weight data in the following manner. Let X be an integer and let the fact that an observation falls between $X - \frac{1}{2}$ and $X + \frac{1}{2}$ be recorded by placing an area centered above X on the axis of real numbers. After six leaves are weighed and recorded, the diagram might appear as shown in Figure 1.1. Diagrams of the type pictured in Figure 1.1 are referred to as *histograms*. If the

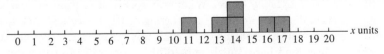

Figure 1.1

[1] Technically, this is a simple random sample.

weight is now measured to the nearest half unit, 16 leaves are examined
instead of 6, and smaller areas are used to record volumes, the histogram
might appear as shown in Figure 1.2. If more delicate instruments are

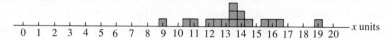

Figure 1.2

employed, the weight is measured to the nearest quarter unit, 160 leaves
are examined, and the areas used to depict weight on the graph are
made very small, the histogram might appear as shown in Figure 1.3.

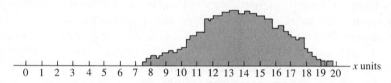

Figure 1.3

Conceive now of the theoretical situation where every leaf on a large tree
(there are a million or more leaves on some large trees) has its weight
measured accurately to several decimal places and these weights are
recorded with very small areas in the manner described above. If, in
addition, the entire area pictured in the histogram is forced to be exactly
one square unit, then the situation may be pictured as shown in Figure 1.4.

Figure 1.4

A histogram such as this is called a population *density* or a population
frequency distribution. The boundary of the histogram is referred to as a
density curve or *frequency curve*. When we speak of a random variable
having a certain density, we shall be referring to its theoretical histogram.
In many experimental situations, we shall not create a density curve in
the manner described here, but it is helpful to conceive of such a density
in association with a random variable and the concept is indeed an
important one.

Pursuing the illustration further, suppose that a measure is required for the plausibility of the following statement: *The next leaf picked at random from the tree will have a weight less than twelve units.* If the only datum available for supplying an answer to the plausibility of this statement is the histogram related to the six observations, we might reason as follows. One of the six leaves studied had weight less than 12 units, hence the relative frequency of such events in the past was $\frac{1}{6}$; this number could serve as our measure of plausibility. Applying the same reasoning to information available in the form of the second histogram, we might conclude that since 3.5 units of the total 16 units of area lie to the left of 12, then 3.5/16 is a proper measure of the plausibility of the statement. Carrying this reasoning to the situation where the population density function is available, we say that the area under the curve to the left of 12 would serve as a measure of the plausibility of the statement. This type of plausibility measure is called probability.

1.8 AN ELEMENTARY DISCUSSION OF PROBABILITY AND ITS PROPERTIES

Mathematics is associated with symbols, not so much because they are necessary but because they are convenient. If notation is correctly understood by the reader, it allows the writer to say in one line what would require a paragraph if written out in words. The student should be aware and keep in mind that notational facts are more a part of language than science. We can no more expect to understand mathematics without learning the notation than we can expect to understand Russian without learning the letter symbols. The mathematical aspects of probability treated here will be found easy, if the notation is studied and understood.

A *probability* is a number attached to an event. Events will be denoted by capital letters: A, B, and so on. Events may be thought of as outcomes of an operation or experiment. Previously, we considered the event that the next leaf picked from a tree will have weight less than 12 units. $P(A)$ will denote the probability attached to the event A. For a general class of events there has been given no universally acceptable elementary definition for probability, but for events where a random variable falls in an interval and the density curve is available, the definition in terms of area under the curve is quite satisfactory. In particular, the student is urged to verify that probability defined this way satisfies the following intuitive properties that mathematicians agree must be satisfied by a definition of probability:

(a) $0 \le P(A) \le 1$ for all events A.
(b)   If A can not happen, $P(A) = 0$.
(c)   If A is sure to happen, $P(A) = 1$.
(d)   If A_1 and A_2 are *mutually exclusive* (that is, they cannot both happen), then

$$P(A_1 \text{ or } A_2) = P(A_1) + P(A_2).$$

We now focus our attention on a particular density curve, which is the *standardized normal curve*. Many properties of this curve will be studied in detail later. At present, we shall proceed to determine probabilities for events associated with this density. The standardized normal curve is pictured in Figure 1.5. Consider the problem of determining $P(A)$,

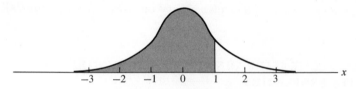

Figure 1.5   The Standard Normal Curve.

where A is the event that $X < 1$ and X is the standardized normal variable. Applying the definition $P(A) = P(X < 1)$ = area under the curve to the left of one, the required area is the shaded area in Figure 1.5. Remembering that the entire area under the curve is one unit, the required probability appears to be near .85. The actual computation of this and other areas under curves is not generally an elementary problem, but many areas have been tabulated in statistical tables. The particular table needed here is Table I. Reading from this table we obtain, as an answer to our problem, $P[X < 1] = .8413$.

In many situations, it becomes important to assign probabilities to an event expressed in terms of other events. At this time, we shall study three important special cases. $A_1 \cup A_2$ referred to as the *union* of A_1 and A_2 is the event that either A_1 or A_2 or both A_1 and A_2 happen. $A_1 A_2$ denotes the event that both A_1 and A_2 occur and is called the *intersection* of A_1 and A_2. The *complement* of A_1 is denoted by $\bar{A}_1$ and is the event that A_1 does not happen. Probabilities assigned to these events will be illustrated by again considering the standardized normal curve. Let A_1 be the event that $X > 1.2$ and let A_2 be the event that $0 < X < 2.4$. The probability assigned to $A_1 \cup A_2$ is pictured in Figure 1.6, the probability assigned to $A_1 A_2$ is shown in Figure 1.7, and the probability

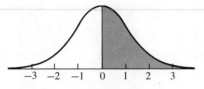

Figure 1.6

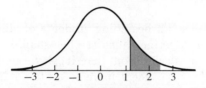

Figure 1.7

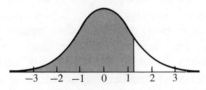

Figure 1.8

attached to $\bar{A}_1$ is displayed in Figure 1.8. The actual values of these probabilities are left for the student to determine with the aid of Table I. Two important, although almost obvious, probability rules are illustrated in Figures 1.6 to 1.8. The first, known as the *addition rule* is:

$$P(A_1 \cup A_2) = P(A_1) + P(A_2) - P(A_1 A_2).$$

The second, known as the *complement rule* is:

$$P(\bar{A}_1) = 1 - P(A_1).$$

If $A_1 A_2$ cannot happen, the intersection is called *null* and $P(A_1 A_2) = 0$. This, of course, is another way of saying that A_1 and A_2 are mutually exclusive, in which case the addition rule reduces to

$$P(A_1 \cup A_2) = P(A_1) + P(A_2).$$

1.9   EXERCISES INVOLVING THE STANDARDIZED NORMAL TABLE

Suppose that the random variable X has a standardized normal distribution. Using Table I, obtain the following probabilities:

(a) $P[X < -1]$ (b) $P[X < 2.5]$
(c) $P[X < 1.645]$ (d) $P[1.5 < X]$
(e) $P[-1 < X < 1]$ (f) $P[-1.96 < X < 1.96]$
(g)   Verify the addition rule when $A_1 \equiv [-1 < X < 1]$ and $A_2 \equiv [X < \frac{1}{2}]$.

1.10   THE NORMAL CURVE

The fact that almost all beginning students of statistics have heard about the normal curve, often called the bell-shaped or Gaussian curve, gives evidence to the declaration that it plays an important role in science. The *normal curve* concept more accurately relates to a family of curves, one of which is the standardized normal curve. Many "real-world" distributions are described adequately by a member of the normal family. One important statistical problem is to determine which member of the normal family best describes the "real-world" distribution. Biologists inform us that leaves on a tree have a weight distribution adequately described by a normal curve. Other familiar examples are heights of males (females), intelligence quotients of Americans, weight gains per day for newborn pigs of a particular breed, and times required to perform a specific industrial operation. Another reason why the normal curve plays an important role in statistics is that it lends itself to mathematical analysis. Many problems have been solved only for situations where normality has been assumed. Other important reasons for studying normal random variables will come to light later.

A particular member of the family of normal curves is uniquely determined by the values of two parameters. These are the population mean and the population variance denoted, respectively, by μ and σ^2. Another way of saying the same thing is the following: If a random variable X is known to possess a normal density, then knowledge of the values of μ and σ^2 determines uniquely the density curve for X. Two normal curves appear in Figure 1.9. Normal curves are symmetric about the mean of

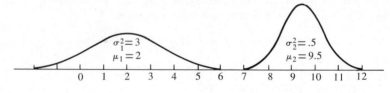

Figure 1.9

the random variable with which they are associated and, being density curves, they of course cover one unit of area. The range on all normal random variables is from negative infinity to positive infinity. This is one reason why "real-world" random variables are never exactly characterized by a normal density.

Areas under curves again play the role of probabilities. It is clear, however, that we could not possibly tabulate areas for every combination of μ and σ^2. Defining probability to be the area under a curve would be of little value were it not for the very important fact that all normal random variables can easily be transformed to the tabulated standard normal variable. Before stating the pertinent facts concerning the transformations, we shall introduce notation that will make the statement of these facts more concise and, hopefully, easier to understand. By $X \sim N(\mu, \sigma^2)$ we mean that the random variable X is distributed normally with mean μ and variance σ^2. The *standardized normal* random variable, say Z, is the particular normal variable that has mean zero and variance one. In our notation, $Z \sim N(0, 1)$. Because of its importance, we make our statement in the form of a theorem.

Theorem 1.2 *If $X \sim N(\mu, \sigma^2)$, then $b(X \pm a) \sim N[b(\mu \pm a), b^2\sigma^2]$.*
Important corollaries are the following:

   (1)   If $X \sim N(\mu, \sigma^2)$, then $(X - \mu)/\sigma \sim N(0, 1)$.
   (2)   If $X \sim N(\mu, \sigma^2)$, then $(X - \mu) \sim N(0, \sigma^2)$.
   (3)   If $X \sim N(\mu, \sigma^2)$, then $(X/\sigma) \sim N(\mu, 1)$.

This theorem, like many of the facts to follow, will not be proved, but we will discuss its meaning in order to give insight into its importance. Notice first that a similar relationship holds for sample means and sample variances; that is, if $X_i = b(Y_i \pm a)$, $i = 1, \ldots, n$, then $\bar{X} = b(\bar{Y} \pm a)$ and $S^2{}_X = b^2 S^2{}_Y$. These facts the reader can establish from the definition of sample mean and the definition of sample variance. A special case of these results was considered in Section 1.5. Since μ is the population counterpart of $\bar{X}$ and σ^2 is the population counterpart of S^2, it seems reasonable that the population mean of $b(X \pm a)$ should be $b(\mu \pm a)$ and that the population variance of $b(X \pm a)$ should be $b^2\sigma^2$. Variance is not affected by adding or subtracting a constant, for the operation merely moves the entire distribution intact and centers it at the new mean. The preserving of normality when a constant is added to or subtracted from a random variable is quite apparent from a consideration of the density curve (see Figure 1.10), but the fact that bX is normal when X is normal is perhaps less easy to argue with pictures.

Theorem 1.2 provides information necessary to obtain probabilities for events associated with a normal curve, provided that we know μ and

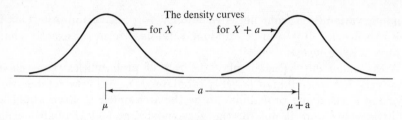

The density curves
for X for $X + a$

μ $\mu + a$

Figure 1.10

σ^2. Consider the problem of obtaining $P[X < 3]$ when $X \sim N(2, 9)$. Applying Theorem 1.2, we can write

$$P[X < 3] = P\left[\frac{X - \mu}{\sigma} < \frac{3 - \mu}{\sigma}\right] = P\left[\frac{X - 2}{3} < \frac{3 - 2}{3}\right] = P\left[Z < \frac{1}{3}\right]$$

where $Z = (X - 2)/3$ is a standardized normal variable. Since the density of Z is tabulated, from Table I we read $P[X < 3] = .6295$.

1.11 EXERCISES INVOLVING NORMAL VARIABLES

(1)   Compute $P[X < 4]$ if $X \sim N(2, 100)$.

(2)   Compute $P[X > 5]$ if $X \sim N(6, 16)$.

(3)   Compute $P[10 < X < 20]$ if $X \sim N(12, 9)$.

(4)   Compute $P[-.71 < X < 1.25]$ if $X \sim N(-.54, 4)$.

(5)   Compute $P[X > .38]$ if $X \sim N(.505, 9)$.

(6)   Compute $P[.182 < X < 2]$ if $X \sim N(-1.2, 1.21)$.

(7)   Compute $P\left[\left|\frac{X - \mu}{\sigma}\right| > 1.96\right]$ if $X \sim N(\mu, \sigma^2)$.

(8)   What is a if $P\left[\left|\frac{X - \mu}{\sigma}\right| > a\right] = .1$ and $X \sim N(\mu, \sigma^2)$?

(9)   What is a if $P\left[-a < \frac{X - \mu}{\sigma} < a\right] = .95$ and $X \sim N(\mu, \sigma^2)$?

(10)   What is μ if $P[X > 8] = .5$ and $X \sim N(\mu, \sigma^2)$?

1.12 DEGREES OF FREEDOM

Degrees of freedom is another of the basic concepts used frequently in this text. The idea is always the same, but it appears in what may at first seem like totally different contexts. Consider the sum of deviations $\sum_{i=1}^{n} (X_i - \bar{X})$ which we have seen always equals zero. Let the deviations $X_i - \bar{X}$ be denoted by d_i and suppose that we play a game, the object of which is to create deviations d_i such that $\sum_{i=1}^{n} d_i = 0$. We are free to choose any of the $(n - 1)$ d_i's but the remaining deviation d_r is restricted to be $d_r = - \sum_{i \neq r}^{n} d_i$. For example, if $n = 4$, and we arbitrarily choose $d_1 = 10$, $d_2 = -5$, and $d_4 = -2$, then we have no freedom to select d_3, for d_3 must be -3. We consequently say that there are three degrees of freedom in our game, when $n = 4$. In general, $\sum_{i=1}^{n} (X_i - \bar{X})$ has $(n - 1)$ degrees of freedom.

The game can also be played in two dimensions. Consider a two-way rectangular layout of numbers such as the matrix in Figure 1.11. The totals for each row and each column are called marginal totals.

			Totals
2	5	4	11
7	4	11	22
Totals 9	9	15	33

Figure 1.11

Consider now a two-way layout with marginal total only, such as we find in Figure 1.12. If two numbers a and b are selected for two cells, not in the same column, then the remaining cells are determined as

			Totals
			20
			30
Totals 10	15	25	50

Figure 1.12

			Totals
a	b	$20-a-b$	20
$10-a$	$15-b$	$5+a+b$	30
Totals 10	15	25	50

Figure 1.13

shown in Figure 1.13. We are free to choose but two cells and thus the game is said to have two degrees of freedom.

The reader can verify that this game played with a two-way layout involving r rows and c columns has $(r - 1)(c - 1)$ degrees of freedom.

If the sample variance for a sample of size n is held constant, there are many sets of d_i's which will satisfy the equation

$$(n - 1)S^2{}_x = \sum_{i=1}^{n} (X_i - \bar{X})^2 = \sum_{i=1}^{n} d_i{}^2.$$

In each of these sets, one number, say d_r, must be chosen so that both $d_r = - \sum_{i \neq r} d_i$ and $d^2{}_r = (n - 1)S^2{}_x - \sum_{i \neq r} d^2{}_i$. This is one reason why $(n - 1)$ degrees of freedom is associated with $\sum_{i=1}^{n} (X_i - \bar{X})^2$, and in part explains the use of $(n - 1)$ instead of n in the definition of S^2.

Consider four autos at a road intersection. Suppose that it is agreed that the autos each should go in a different direction. The first three autos to leave have choices but the last auto does not. We have here three autos with freedom of choice.

2

STATISTICAL INFERENCE RELATIVE TO SAMPLING FROM ONE NORMAL POPULATION

2.1 INTRODUCTION TO SAMPLING

Much of elementary statistics concerns itself with making statements about the mean μ and the variance σ^2 of a population. When statements relative to the population are made from information based on a sample from the population, *statistical inference* is in operation. The exact values of μ and σ^2 generally cannot be obtained from a sample, but with proper methods of selecting the sample, μ and σ^2 can be estimated in such a way that the nature of the approximation can be described.

In this chapter, we discuss samples where the observations are those taken on by *independent* random variables. Two random variables are said to be independent if the selection of the value taken on by the second variable is in no way influenced by the value taken on by the first variable. The generalization of this definition goes as follows: k random variables are independent if the selection of the values taken on by j variables $(j = 1, 2, \ldots, k - 1)$ is in no way influenced by the values taken on by the other $(k - j)$ variables. An equivalent statement in terms of probabilities is: The variables $X_1, \ldots, X_k$ are independent if, for every set of real numbers $x_1, \ldots, x_k$,

$$P[X_1 \leq x_1, X_2 \leq x_2, \ldots, X_k \leq x_k] = P[X_1 \leq x_1] \\ \cdot P[X_2 \leq x_2] \cdots P[X_k \leq x_k].$$

A sample of independent observations, in some physical situations, may not be the easiest or preferred type of sample to obtain, but it is the basis for our discussion. It should be emphasized that the methods used to estimate parameters depend on the methods used to obtain the sample.

A random sample from a finite population, as defined in Chapter 1, in general does not provide us with independent observations, but when the population size p is large and the sample size k is small, a random sample does provide us with approximately independent random variable values and the approximation improves as the ratio p/k increases. When the theoretical population is infinite such as the normal is, then random sampling is equivalent to obtaining independent observations.

Suppose then that we have independent normal observations X_1, X_2, . . . , X_n. A shorthand notation for this situation is to write $X_i \sim NID(\mu, \sigma^2)$ which is read: "The X_i are distributed normally and independently with mean equal to μ and variance equal to σ^2." In such a situation, estimators of μ and σ^2 with many good properties are $\bar{X}$ and S^2. The random variables $\bar{X}$ and S^2 are referred to as point estimators and we often write $\hat{\mu} = \bar{X}$ and $\hat{\sigma}^2 = S^2$ to indicate that $\bar{X}$ is an estimator of μ and S^2 is an estimator of σ^2. It has become conventional to use Greek letters for unknown parameters and English letters for their estimators or sample counterparts.

2.2 THE NATURE OF $\bar{X}$

Of the many important properties possessed by the estimator $\bar{X}$, we mention here that it is easily computed and that $\bar{X}$ is an *unbiased* estimator of μ. The general meaning of the word *"unbiased"* will be considered in detail later but its role with reference to $\bar{X}$ will be noted in the present discussion.

The fact that $\bar{X}$ is a variable needs to be emphasized and contemplated. Consider samples taken at random from a large population. For each sample, focus upon $\bar{X}$. The value of $\bar{X}$ in general will not be the same for all samples; that is, the random variable $\bar{X}$ possesses a distribution. For normal populations, we have the following remarkable theorem.

Theorem 2.1 *If* $X_i \sim NID(\mu, \sigma^2)$, *then* $\bar{X} \sim N(\mu, \sigma^2/n)$.

Much of statistical inference rests upon this theorem and its generalizations. The theorem combines three very important facts. First, it says that if $\{X_1, . . . , X_n\}$ is a random sample from a normal population, then the average $\bar{X}$ is a normal random variable. This, to the discerning reader, is a rather remarkable fact. It says that, if we sample at random from a member of the normal family of distributions, then the average is a variable that belongs to the same family. This is not the case for many families of distributions.

The second great truth expressed in this theorem is that when each X_i has mean μ, then likewise the mean of $\bar{X}$ is μ. We refer to $\bar{X}$ as an unbiased estimator of μ because of this fact. The requirement that the X_i be normal is irrelevant. In everyday language, the statement merely expresses the thought that we expect the same from an average as we do from individual observations.

The third fact found recorded in this theorem is generalized in Theorem 2.2. This fact will be emphasized to an extent surpassed only by the emphasis given to one other statement in this text.

Theorem 2.2 *If $X_1 \cdots X_n$ is a random sample then THE VARI-ANCE OF ANY OBSERVATION X_i IS n TIMES THE VARIANCE OF $\bar{X}$.* In symbols,

$$\sigma^2_{\bar{x}} = \frac{\sigma^2_x}{n}.$$

The fact that people repeat experiments, form committees to make decisions, consider evidence more than once, make several trials, ask the same question of several experts, make duplicate observations, measure quantities twice, and the like, is outward manifestation of the principle embodied in the above theorem. The theorem elaborates on the truth that averages have less variability than do individual observations.

No attempt will be made to formally prove Theorem 2.1 or Theorem 2.2 but an effort will be made to cement the ideas by considering the sample estimates of the parameters involved. Accepting the stated fact that

$$S^2_x = \frac{1}{n-1} \sum_{i=1}^{n} (X_i - \bar{X})^2$$

is a good estimator of σ^2_x implies the acceptance of the fact that S^2_x/n is a good estimator of σ^2_x/n. We then are led to the following definition:

$$S^2_{\bar{x}} = \frac{S^2_x}{n}.$$

The concept presented here is important; hence, to make more sure that its meaning has been properly conveyed to the student, the idea is expressed again in the following words. If the variance of the random variable $\bar{X}$ is denoted by $\sigma^2{}_{\bar{x}}$ and this parameter is to be estimated, then a good estimator of $\sigma^2{}_{\bar{x}}$ is $S^2{}_X/n$ and this statistic will be denoted by the suggestive notation $S^2{}_{\bar{X}}$.

Another procedure for estimating $\sigma^2{}_{\bar{x}}$, which at first may seem more natural, is to use the sample variance formula in conjunction with a sample of $\bar{X}$'s. Since $S_{\bar{x}}{}^2$ has already been used, we must select another symbol for this estimator and $S^{*2}{}_{\bar{x}}$ is the notation used here. Consider k different means $\bar{X}_1, \ldots, \bar{X}_k$ computed from $k > 1$ samples, each of size n, then

$$S^{*2}{}_{\bar{x}} = \frac{1}{k-1} \sum_{i=1}^{k} (\bar{X}_i - \bar{\bar{X}})^2 \qquad \text{where} \qquad \bar{\bar{X}} = \frac{1}{k} \sum_{i=1}^{k} \bar{X}_i.$$

$S^2{}_{\bar{x}}$ and $S^{*2}{}_{\bar{x}}$ are estimates of the same parameter, $\sigma^2{}_{\bar{x}}$, but being random variables they may take on, in any particular situation, quite different values. In actual practice we often obtain but one sample, in which case $\sigma^2{}_{\bar{x}}$ can be estimated by $S^2{}_{\bar{x}}$ but $S^{*2}{}_{\bar{x}}$ is meaningless.

The student of statistics should study the illustrative example that follows, then work the computational problems and think about their relevance to the formulas and theorems presented here. Thinking about exercises of this type is perhaps the best means for providing insight into the nature of variability and normality.

2.3   AN EXAMPLE TO ILLUSTRATE THE VARIABILITY OF $\bar{X}$

Consider groups of five senior high-school boys who appear for military examinations. The word "group" will here be synonymous with the word "sample." Suppose that the numbers recorded in Table 2.1 are the test scores for five boys in each of seven groups selected at random from many groups that appeared for the examination.

Table 2.1   Test Scores for Seven Groups of Boys

Group I	II	III	IV	V	VI	VII
74	86	72	73	83	69	80
68	62	90	78	88	75	65
85	80	74	85	80	79	87
79	73	79	64	79	80	79
63	66	80	83	72	90	71
$\bar{Y}$ 73.8	73.4	79.0	76.6	80.4	78.6	76.4
$S^2{}_Y$ 75.7	96.8	49.0	71.3	34.3	59.3	72.8
$S^2{}_{\bar{Y}}$ 15.14	19.36	9.80	14.26	6.86	11.86	14.56

In Table 2.1, we want to think of the scores recorded as being part of a population of scores corresponding to the collection of all senior high-school boys who might have appeared for the military examination. The unknown population variance will be denoted by σ^2. Each group is a part of the population and, if we apply the assumption that each group is a random sample from the population, then $S^2{}_y$ for each group is an estimate of σ^2. Note that the values of $S^2{}_Y$ range from 34.3 to 96.8. This is typical, because estimates based on five observations or less tend to be quite variable and indeed sample variances are generally far more variable than sample means.

In addition to a population of scores, a population of group averages can be contemplated. The seven group averages can be thought of as a sample of size seven from the population of all possible group averages. This population has the same mean as does the population of all scores. Calling the population mean μ, we then have as estimates of μ, $\bar{Y}_1$, . . . , $\bar{Y}_7$, and $\bar{\bar{Y}}$. If the variance of the group averages is denoted by $\sigma^2{}_{\bar{Y}}$, then according to Theorem 2.2, $\sigma^2{}_{\bar{Y}} = \sigma^2{}_Y/5$. From each group an estimate of $\sigma^2{}_{\bar{Y}}$ was computed. $S_{\bar{Y}}{}^2$ ranged from 6.86 to 19.36. Notice that as the theory indicates, the group averages $\bar{Y}_1$, . . . , $\bar{Y}_7$ are not nearly as variable as the individual observations. We emphasize again that the variance of group averages decreases as the sample size increases.

From the sample of seven group means, we can compute another estimate of $\sigma^2{}_{\bar{Y}}$, that is,

$$S^{*2}{}_{\bar{Y}} = \frac{1}{6} \sum_{i=1}^{7} (\bar{Y}_i - \bar{\bar{Y}})^2.$$

The data yield $\bar{\bar{Y}} = 76.886$ and $S^{*2}{}_{\bar{Y}} = 6.961$. A study of the relative merits of these estimators of $\sigma^2{}_{\bar{Y}}$ and others, will be delayed. It will suffice here to say that in practice, often one sample is taken from a population, in which case $S^2{}_{\bar{Y}}$ is the estimator used.

2.4 COMPUTATIONAL EXERCISES RELATED TO THE DISTRIBUTION OF $\bar{X}$

(1) For each of the k different samples of size six taken in connection with Exercise (7), Section 1.6, compute $\bar{Y}_i$ and $S^2{}_{\bar{Y}}$, $i = 1, 2, . . . , k$.

(2) Compute

$$\bar{\bar{Y}} = \frac{1}{k} \sum_{i=1}^{k} \bar{Y}_i$$

and

$$S^{*2}_{\bar{Y}} = \frac{1}{k-1}\left[\sum_{i=1}^{k}\bar{Y}_i^2 - \frac{1}{k}\left(\sum_{i=1}^{k}\bar{Y}_i\right)^2\right]$$

from the data in Exercise (1). Compare $S^{*2}_{\bar{Y}}$ and the $S^2_{\bar{Y}_i}$ remembering that they all estimate $\sigma^2_{\bar{Y}}$.

(3) From the following data estimate μ, σ^2_X and $\sigma^2_{\bar{X}}$ where $\bar{X} = \frac{1}{8}\sum_{i=1}^{8}X_i$.

$X_1 = 2$, $X_2 = 4$, $X_3 = 3$, $X_4 = 5$, $X_5 = 4$, $X_6 = 5$, $X_7 = 4$, $X_8 = 3$.

(4) Suppose that $Y_1 = 25$, $Y_2 = 39$, $Y_3 = 31$, $Y_4 = 35$, and $Y_5 = 45$. Consider these observations to be a sample from a population where $Y_i \sim \text{NID}(\mu, \sigma^2)$. State the distribution of the random variable $\bar{Y} = \frac{1}{3}\sum_{i=1}^{3}Y_i$. From the above five observations, estimate $\sigma^2_{\bar{Y}}$, where $\bar{Y}$ is the average of three Y's.

(5) Estimate the mean and the variance of $\bar{X} = \frac{1}{9}\sum_{i=1}^{9}X_i$ if the following data are the result of a random sample of X_i's.

$X_1 = 13.9$	$X_4 = 14.8$	$X_7 = 15.3$
$X_2 = 14.2$	$X_5 = 13.3$	$X_8 = 14.3$
$X_3 = 15.0$	$X_6 = 14.5$	$X_9 = 14.7$

(6) Given a random sample of size 10 from a normal population with mean $\mu = 2$ and variance $\sigma^2 = 40$ find:
(a) $P(\bar{X} \leq 3)$ (b) $P(\bar{X} \leq .8)$
(c) $P(\bar{X} \geq 1)$ (d) $P(1.5 \leq \bar{X} \leq 2.7)$
(e) $P(|\bar{X}| \leq 1)$ (f) $P(-1.2 \leq \bar{X} \leq .5)$

(7) Given a random sample of size 25 from a normal population with mean 100 and variance 225 find:
(a) $P[\bar{X} < 106]$ (b) $P[97 < \bar{X} < 103]$
(c) $P[101 < \bar{X} < 105]$ (d) $P[|\bar{X} - 100| > 6]$

(8) About how many observations are required so that, when sampling from a normal population with mean $\mu = 0$ and variance $\sigma^2 = 4$, we shall have $P(-1 \leq \bar{X} \leq 1) = .95$?

2.5 A SAMPLE-SIZE PROBLEM

One of the questions most often asked of statisticians by experimenters is: "How large a sample must I take in order that . . . ." Generally the experimenter is satisfied with a rough sort of answer. This is fortunate for the statistician, because most *sample-size* problems are rather difficult to state precisely, some are difficult to solve, and those that can be stated precisely and then solved often have precise answers which are difficult to evaluate in a specific case, unless the aid of a computing machine is obtained. Since many answers to sample-size questions are guesses, the subject should be studied in order to obtain a feeling for the problem.

One of the few relatively easy *sample-size* problems will now be discussed. Consider the problem of determining the sample size necessary in order that the probability that $\bar{X}$ differs from μ by less than k units be .95 when the population sampled is normal with mean μ and variance σ^2. In symbols, we desire n such that $P[-k < \bar{X} - \mu < k] = .95$. Now $\bar{X} \sim N(\mu, \sigma^2/n)$ and

$$\frac{\bar{X} - \mu}{\sigma/\sqrt{n}} \sim N(0, 1);$$

hence, an equivalent statement of the problem is to ask for n, such that

$$P\left(\frac{\bar{X} - \mu}{\sigma/\sqrt{n}} < \frac{k\sqrt{n}}{\sigma}\right) = .975.$$

Setting $(k\sqrt{n})/\sigma = 1.96$, we have $n = [(1.96)/k]^2$. This is purely an academic answer since, in general, this is not an integer. We can say, however, that when n is greater than $[(1.96\sigma)/k]^2$ we have $P[-k < \bar{X} - \mu < k] \geq .95$.

2.6   EXERCISES IN DETERMINING APPROXIMATE SAMPLE SIZES

Determine the approximate sample size needed, in order that the probability that $\bar{X}$ differs from μ by

(1)   less than 3 units be about .95 if sampling is done from a normal population with variance equal to 36,

(2)   more than 2 units be less than .05 if sampling is done from a normal population with variance equal to 25,

(3)   more than $\frac{1}{2}$ unit, be less than .05 if sampling is done from a normal population with variance equal to 9, and

(4)   less than 5 units, be greater than .99 if sampling is done from a normal population with variance equal to 64.

2.7   THE CONCEPT OF AN INTERVAL ESTIMATE

Many experimenters report their findings in the form of an interval statement. A physicist may report a pressure to be $30 \pm \frac{1}{2}$ pounds per square inch. An astronomer may report a heavenly body between 80 and 81 light years away. In like manner, statistical facts in all disciplines can be given in an interval form. *The concept of an interval estimate may be made more refined by stating, in addition to the interval, a probability associated with the procedure for obtaining the interval.* When the probability associated with the procedure is multiplied by 100, it becomes a percentage used as a measure of confidence attached to the procedure for generating intervals, and any interval produced by the procedure is referred to as a confidence interval. To be specific, early remarks will be confined to 95 percent confidence intervals.

At this point in our development of statistics, we are prepared to consider only one confidence interval situation, that being the case where the interval is for μ and it is based on a random sample of size n from a normal population with known variance. This is not a realistic problem but we study it because the objective now is to convey the concept of a 95 percent confidence interval.

One procedure for generating 95 percent confidence intervals is to take random samples of size n from the normal population under consideration and compute the intervals $\{\bar{X} \pm 1.96\sigma/\sqrt{n}\}$. A typical collection of intervals computed in this manner is pictured in Figure 2.1. This proce-

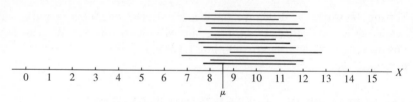

Figure 2.1

dure has the property that 95 percent of the intervals generated, after infinite application of the procedure, include (or cover) the parameter μ. The validity of the preceding statements will be proved after we take notice of the following facts: (1) The midpoint of each interval computed by the above procedure is $\bar{X}$. (2) After we have computed a confidence interval, we still do not know if it contains the parameter μ, for indeed the probability for this event can be made 1 only by selecting the interval $-\infty$ to ∞. (3) The length of a 95 percent confidence interval for μ when σ^2 is known is $2[(1.96)\sigma]/\sqrt{n}$. Thus, if samples of size n are taken repeatedly from the same normal population, the interval estimates have the same length but differ in location.

To prove that the procedure has the 95 percent covering property mentioned earlier, we begin by recalling that

$$\left[\frac{\bar{X} - \mu}{\sigma/\sqrt{n}}\right] \sim N(0, 1).$$

Hence,

$$P\left[-1.96 < \frac{\bar{X} - \mu}{\sigma/\sqrt{n}} < 1.96\right] = .95$$

then

$$P\left[\frac{-1.96\sigma}{\sqrt{n}} < \bar{X} - \mu < \frac{1.96\sigma}{\sqrt{n}}\right] = .95$$

$$P\left[\frac{-1.96\sigma}{\sqrt{n}} < \mu - \bar{X} < \frac{1.96\sigma}{\sqrt{n}}\right] = .95$$

and finally

$$P\left[\bar{X} - \frac{1.96\sigma}{\sqrt{n}} < \mu < \bar{X} + \frac{1.96\sigma}{\sqrt{n}}\right] = .95.$$

The procedure when a $100p$ percent confidence interval is called for gets changed only in that 1.96 is replaced by the appropriate number read from the normal table. Thus, if a 99 percent confidence interval is desired, we replace 1.96 by 2.576. The more realistic situation where σ^2 is unknown is handled by using s^2 as an estimate of σ^2 and replacing 1.96 by the appropriate number, depending on n, which is read from a table called "Student's table." This problem will be discussed after some small sample theory has been introduced in a later section.

2.8   EXERCISES INVOLVING INTERVAL ESTIMATES OF μ WHEN δ² IS KNOWN

(1) Compute 95 percent confidence intervals for μ in the following normal situations. In each case compute the length of the interval.
   (a) $\sigma^2 = 25$, $X_1 = 2.7$, $X_2 = 9.3$, $X_3 = 5.8$, $X_4 = 10.2$;
   (b) $\sigma^2 = 25$, $X_1 = 19$, $X_2 = 10$, $X_3 = 7$, $X_4 = 20$;
   (c) $\sigma^2 = 25$, $n = 4$, $\bar{X} = 104$;
   (d) $\sigma^2 = 25$, $n = 100$, $\bar{X} = 106$;
   (e) $\sigma^2 = 4$, $n = 4$, $\bar{X} = 108$;
   (f) $\sigma^2 = 64$, $n = 400$, $\bar{X} = 52$;
   (g) $\sigma^2 = 1$, $n = 400$, $\bar{X} = 48$.

(2) Compute 99 percent confidence interval estimates of μ for each of the situations in Exercise (1), and compare the lengths of the intervals with those obtained in answer to Exercise (1).

(3) Compute 90 percent confidence intervals for μ in each of the situations described in Exercise (1), then compare the interval lengths with those obtained when 95 percent and 99 percent confidence interval estimates were computed.

(4)   A laboratory-type problem: Consider an industrial process, where for years the variance of the measurements for a certain characteristic of the product had been rather consistently equal to 9, and the measurements were described adequately by a normal density curve. Suppose that the process is changed in such a way that engineers are sure that the variance has not changed significantly from 9 but that the mean has obviously changed. In order to estimate the new mean, normality is assumed and 16 measurements are taken as follows:

$$8.32, \quad 12.76, \quad 4.01, \quad 9.75, \quad 7.32, \quad 6.01, \quad 7.75, \quad 5.20$$
$$7.52, \quad 8.08, \quad 10.56, \quad 7.22, \quad 3.40, \quad 8.44, \quad 5.68, \quad 4.95$$

Compute 90 percent, 95 percent, and 99 percent confidence intervals for the new mean and compare the lengths of the confidence intervals.

2.9   THE CONCEPT OF TESTING A STATISTICAL HYPOTHESIS

A *statistical hypothesis* is any assumption relative to the distribution of random variables and a *test of a statistical hypothesis* is any procedure for deciding whether to accept or to reject the statistical hypothesis. The discussion, at the present time, will be confined to procedures based on information found in a random sample taken from a population about which the statistical hypothesis was made. Furthermore, we avoid studying general hypothesis-testing problems by applying other restrictions, so that for a particular hypothesis, in a specific situation, we can talk about a procedure which, in some sense, is best.

It is helpful to partition hypothesis-testing problems into categories. The main category from which we shall investigate problems is the category where the hypotheses are expressed in terms of values of parameters. Hypotheses are often stated in such a way that we hope the information gathered from a sample will nullify the hypothesis. This sort of reasoning is similar to that used when we apply mathematical proof by contradiction. Thus, the name *null hypothesis* was coined for the hypothesis under test. Obviously, the situation is very uninteresting if there is no alternative to the null hypothesis. The totality of all "possible" alternatives to the null hypothesis constitutes what we call the *alternative hypothesis*. Not all assumptions made in a situation need be statistical hypotheses subject to testing. For example, a researcher may assume that his observations come at random from a normal population with mean $\mu = 5$ and variance $\sigma^2 = 9$. It may be that only the validity of the statement made concerning μ is of concern. If normality and the value of the variance are accepted as fact, then the alternative hypothesis is $\mu \neq 5$. On the other hand, if only normality is accepted, then the alternative to $\mu = 5$ and $\sigma^2 = 9$ is $\mu \neq 5$ and $\sigma^2 \neq 9$.

Hypotheses can be represented in what will be called *parameter spaces*. Three parameter spaces, associated with the parameters μ and σ^2 for a normal density, are: (1) the real-number line $-\infty$ to ∞ for values of μ; (2) the positive real-number line 0 to ∞ for values of σ^2; and (3) the upper half plane, $-\infty < \mu < \infty, \sigma^2 > 0$, for pairs (μ, σ^2). If a hypothesis can be represented by one point in the parameter space, the hypothesis is called *simple*. When a hypothesis is not simple, we say that it is composite. $\mu = 5$ and $\sigma^2 = 9$ is a simple hypothesis in the space of (μ, σ^2), while $\mu \neq 5$ and $\sigma^2 \neq 9$ is a composite hypothesis.

Hypothesis-testing problems can be partitioned into subcategories characterized by the nature of the null and alternative hypotheses. The state of affairs with respect to null versus alternative can be: (1) simple versus simple, (2) simple versus composite, (3) composite versus simple, and (4) composite versus composite. Possibility (3) is seldom, if ever, studied since, with a change in the name of the hypothesis, it is put into the form of possibility (2), and (2) has proved easier to work with than (3).

We consider first a simple-versus-simple situation. Either hypothesis can be designated the null hypothesis. Two errors may result as the consequence of applying a procedure for deciding whether to accept or to reject the null hypothesis. These errors have the rather standard designations, *Type I* and *Type II*.

A Type I error is the event of rejecting a true null hypothesis.

A Type II error is the event of accepting a false null hypothesis.

The probability of a Type I error will be denoted by α and the probability of a Type II error will be denoted by β.

Consider now the specific normal case where σ^2 is known to be 16, the null hypothesis is $H_0: \mu = 10$, and the alternative hypothesis is $H_1: \mu = 17$. The simplest case, involving a sample size of one, is pictured in Figure 2.2. Without applying a criterion, a principle, or a mathematical theory of any sort, but by merely thinking about the situation, we are intuitively led to reject H_0 if the observation X is near 17 or larger than 17 and to accept H_0 for other values of X. Those values of X for which H_0 is rejected constitute the *critical* (rejection) *region* and the set of points where H_0 is accepted is called the acceptance region. The point c where the acceptance region ends and the critical region begins is referred to as the *critical point*. Figure 2.2 also illustrates the probabilities α and β. Since a Type I error can happen only when H_0 is true, α is the area over the rejection region and under the $N(10, 16)$ curve. In a similar manner, β is pictured as the area over the acceptance region and under the $N(17, 16)$ curve.

Now that the stage has been set, we need to concern ourselves with a

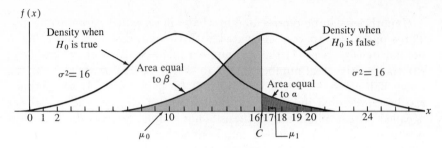

Figure 2.2

theory-type problem and an application-type problem. The first problem concerns how to choose the critical point c and the second problem relates to how we then handle the case where the sample size is greater than one. As we might suspect, the two problems are interrelated.

In considering available procedures for testing a simple null hypothesis against a simple alternative hypothesis, we obviously would select, if possible, a procedure where α and β were both zero and n was small. No procedure based on a random sample exists with these properties and, indeed, for all procedures where α is zero, we have either β equal to one or n infinitely large, and when β is zero either α equals one or n is infinitely large. The problem is somewhat analogous to that faced by a medical doctor as he endeavors to make a decision concerning whether to operate for appendicitis. If the patient has appendicitis and the doctor does not operate an error has been made, and if the patient is operated on and the appendix is found in a good state but the patient grows worse, then another type of error has been committed. The doctor can avoid the first type of error by always operating for appendicitis when he is faced with a sick patient and he can avoid the second type of error by never operating for appendicitis when a patient is sick. Neither of these procedures seems satisfactory.

Realizing that α, β, and n are related, we pursue the task of selecting a procedure (in our example, selecting a critical point c) in such a way that hopefully α, β, and n are acceptable to all concerned with the problem. A convention for simple-versus-simple hypothesis testing situations which yields "good" procedures and which is in common use today is that due to Neyman and Pearson. This convention, which is commonly called the *fundamental principle of hypothesis testing*, states that after selecting α and an appropriate sample size n, then the procedure to be used is that which minimizes β. When this principle was first adopted, it posed an interesting set of problems for mathematical statisticians to work on. The procedures presented in this text are the solutions obtained by mathematical statisticians and we shall refer to them as *best procedures*. It should be pointed out that procedures obtained as solutions to the mathematical problems agreed, for the most part, with those already in use by experimenters.

In accord with the fundamental principle, let α be chosen (say, .05) and let $n = 1$. Turning again to our illustrative example, the critical region starts at the point c where c is such that $P[X > c] = .05$ when H_0 is true. If H_0 is true, $X \sim N(10, 16)$; hence,

$$P[X > c] = P\left[\frac{X - 10}{4} > \frac{c - 10}{4}\right] = .05.$$

Therefore, by Table I, $(c - 10)/4$ must equal 1.645 and $c = 16.58$. In summary, the best test of H_0 based on one observation when α is chosen to be .05 is to reject H_0 if $X \geq 16.58$ and accept H_0 if $X < 16.58$.

A consideration of Figure 2.2 reveals that β is considerably greater than α. The value of β is computed as follows: $\beta = P[X < 16.58]$ where $X \sim N(17, 16)$. Reading from the standardized normal table,

$$\beta = P\left[\frac{X - 17}{4} < \frac{16.58 - 17}{4}\right] = P[Z < -.105] = .4582.$$

In order to reduce β without changing α, a sample size greater than one needs to be considered. Going directly to the general case, let the sample size be n. For a problem such as we are considering, the information contained in n observations $X_1, \ldots, X_n$ can be condensed into one number $\bar{X}$. The best procedure is to find a new critical value $c(n)$, depending on n, such that $P[\bar{X} > c(n)] = .05$ when H_0 is true. The distribution of $\bar{X}$ is $N(10, 16/n)$ when H_0 is true; hence, we seek $c(n)$ such that

$$P\left[\frac{\bar{X} - 10}{4/\sqrt{n}} > \frac{c(n) - 10}{4/\sqrt{n}}\right] = .05.$$

Setting

$$\frac{c(n) - 10}{4/\sqrt{n}} = 1.645$$

we find

$$c(n) = 10 + \frac{6.58}{\sqrt{n}}.$$

A special case will be illustrated by letting $n = 4$. In this case $c(4) = 13.29$ and β which is $P[\bar{X} < 13.29]$, when $\bar{X} \sim N(17, 4)$, takes on the value

$$\beta = P\left[Z < \frac{13.29 - 17}{2}\right] = P[Z < -1.855] = .0318.$$

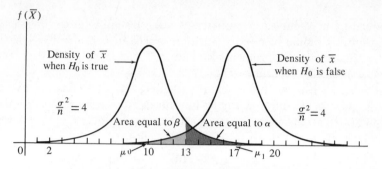

Figure 2.3

This of course is a considerable reduction in the probability of a Type II error. The situation when n is equal to 4 is pictured in Figure 2.3.

2.10 EXERCISES INVOLVING THE TEST OF A SIMPLE HYPOTHESIS ABOUT μ WHEN σ^2 IS KNOWN

For each of the following hypothesis-testing situations
  (a) determine the best critical region,
  (b) determine $\beta = P(\text{Type II error})$, and
  (c) draw an appropriate picture.

(1) Test $H_0: \mu = 8$ versus $H_1: \mu = 14$ when the observations are normal and
  (a) $\alpha = .05$, $n = 1$, and $\sigma^2 = 9$;
  (b) $\alpha = .05$, $n = 4$, and $\sigma^2 = 9$;
  (c) $\alpha = .01$, $n = 4$, and $\sigma^2 = 9$;
  (d) $\alpha = .01$, $n = 9$, and $\sigma^2 = 9$;
  (e) $\alpha = .01$, $n = 9$, and $\sigma^2 = 4$.

(2) Test $H_0: \mu = 15$ versus $H_1: \mu = 13$ when the observations are normal and
  (a) $\alpha = .05$, $n = 6$, and $\sigma^2 = 1.5$;
  (b) $\alpha = .01$, $n = 24$, and $\sigma^2 = 1.5$;
  (c) $\alpha = .01$, $n = 24$, and $\sigma^2 = 6$.

(3) To test $H_0: \mu = 20$ versus $H_1: \mu = 24$ when $\sigma^2 = 49$, an experimenter decides to make the critical point $c = 22$. Since this point is the midpoint of the interval from 20 to 24, α is then forced to equal β when normality is assumed. (Show that this is true by drawing an appropriate picture.) The experimenter wishes to have α and β each less than .05. Show that a sample size of 36 or greater is needed to accomplish this when the population is normal.

(4)   What sample size is needed to test H_0: $\mu = 1.2$ versus H_1: $\mu = 1.6$, when the population is normal with variance equal to 1.44, if the experimenter wants c to be at 1.5 and α less than or equal to .01? Compute the probability of a Type II error after the necessary sample size has been determined.

2.11 STUDENT'S *t* DISTRIBUTION AND ITS ROLE IN STATISTICAL INFERENCE

In inference situations thus far, we have made the unrealistic assumption that the variance of the population was known. In both interval estimation and hypothesis-testing problems, we have been led to consider the standardized normal variable $[(\bar{X} - \mu) \sqrt{n}]/\sigma$. If σ is unknown, then perhaps the natural move is to replace σ by its sample counterpart s. A new family of random variables $[(\bar{X} - \mu) \sqrt{n}]/s$, which depends on the sample size n, must then be considered. As early as 1908, a family of variables closely related to this family was studied by William Gosset who wrote under the name Student. Publishing in the English journal *Biometrika*, in a paper entitled "The Probable Error of a Mean," Gosset set down facts that have dramatically affected the course of statistics. Until Gosset's paper, problems of statistical inference where σ was unknown were handled by replacing σ by s and then ignoring the fact that $[(\bar{X} - \mu) \sqrt{n}]/s$, or an essentially equivalent statistic, was not normal. Gosset and others were aware that this procedure was a good one when n was large, but that it led to incorrect and misleading inferences when the sample size was small. Gosset worked out, or perhaps more correctly put, discovered the distribution curves for a family of random variables much like $[(\bar{X} - \mu) \sqrt{n}]/s$. The problem he solved is now typical of those given beginning students of mathematical statistics. More amazing than the fact that he solved the problem is the way in which he solved it. Today the family of random variables $t(\nu) = [(\bar{X} - \mu) \sqrt{n}]/s$, indexed by $\nu = n - 1$ and computed for random samples from a normal population, is known as *Student's t*. For each value of ν (we recall from a discussion in Chapter 1 that ν is the degrees of freedom associated with s^2), we have a different member of the family of random variables. Attention is directed to Figure 2.4 where density curves for two members of the family are pictured with the standardized normal curve.

Some facts relative to Student's *t* will now be related. The variables have infinite range. The density curves are symmetric about a mean of 0. More area (probability) is found in the tails of Student *t* distributions than in the corresponding tails of the standardized normal distribution. The Student *t* curves approach the standardized normal curve as $\nu =$

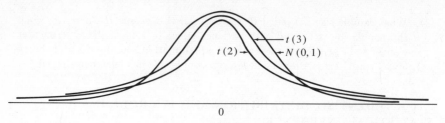

Figure 2.4

$(n - 1)$ becomes large. Areas under Student curves are tabulated in Table III.

The reader can empirically create a histogram of the Student t type by repeatedly computing $[(\bar{X} - \mu) \sqrt{n}]/s$ from random samples of size n taken from a normal population with known mean μ. The Student t family of populations is one of three very important families of populations derived from a parent normal population. The other families that we shall soon investigate are the Chi Square and the Snedecor F families.

Turning attention once again to applications, the situation can be summarized somewhat roughly and certainly briefly by saying that the Student t table is used when σ^2 is unknown in the same way that the standardized normal table is used when σ^2 is known. Taking a more careful look at 100γ percent interval estimation of μ based on a random sample from $N(\mu, \sigma^2)$, we start out by reading the appropriate tabulated values $-t(\nu, \gamma)$ and $t(\nu, \gamma)$ so that

$$P\left[-t(\nu, \gamma) \leq \frac{(\bar{X} - \mu) \sqrt{n}}{s} \leq t(\nu, \gamma) \right] = \gamma$$

and end up with the confidence interval statement

$$P\left[\bar{X} - \frac{t(\nu, \gamma)s}{\sqrt{n}} \leq \mu \leq \bar{X} + \frac{t(\nu, \gamma)s}{\sqrt{n}} \right] = \gamma.$$

Notice that the length of the interval is variable even in situations where n and γ are fixed whereas, when σ^2 is known, the intervals for fixed n and γ have constant length.

The situation with regard to hypothesis testing will be handled by considering the general situation rather than a specific case. Suppose that we wish to test $H_0: \mu = \mu_0$ versus $H_1: \mu = \mu_1$ where $\mu_0 < \mu_1$. Let us base our test on a random sample $X_1, X_2, \ldots, X_n$ from $N(\mu, \sigma^2)$ with σ^2 unknown. Applying the fundamental principle, we select α and then make ourselves aware of the fact that β is minimized if we choose for our critical region those values of $[(\bar{X} - \mu_0) \sqrt{n}]/s$ which exceed t_0, where t_0 is a tabulated number depending on α and ν with the property that $P[t(\nu) > t_0] = \alpha$. If the computed value of $t(\nu) = [(\bar{X} - \mu_0)\sqrt{n}]/s$ is greater than the tabulated value of t_0, we reject H_0, and if $t(\nu)$ is less than t_0, we accept H_0. Although this procedure may seem entirely different from the procedure for handling simple-versus-simple hypothesis-testing situation when σ^2 is known, upon careful investigation it will be found that this is not really the case. If σ^2 is known, we reject H_0 when $\bar{X}$ exceeds the critical value c. This is equivalent to rejecting H_0 when $Z = [(\bar{X} - \mu_0)\sqrt{n}]/\sigma$ exceeds $Z_0 = [(c - \mu)\sqrt{n}]/\sigma$. When we work in the sample space of Z, we use Z_0. Unless we are told to work in the sample space of $\bar{X}$, we shall work henceforth in the sample space of Z when σ^2 is given, and the sample space of t when σ^2 is unknown. Variables such as $Z = [(\bar{X} - \mu_0)\sqrt{n}]/\sigma$ and $t = [(\bar{X} - \mu_0)\sqrt{n}]/s$, when used in conjunction with hypothesis-testing situations, will be referred to as *test statistics*.

2.12   EXERCISES INVOLVING STUDENT'S TABLE

(1)   Compute 95 percent confidence interval for μ in the following normal situations. In each case compute the length of the interval.

(a) $X_1 = 6,$ $X_2 = 8,$ $X_3 = 7;$

(b) $X_1 = 2.7,$ $X_2 = 9.3,$ $X_3 = 5.8,$ $X_4 = 10.2;$

(c) $X_1 = 19,$ $X_2 = 10,$ $X_3 = 7,$ $X_4 = 20;$

(d) $n = 8,$ $\bar{X} = 13.8,$ $s^2 = 2;$

(e) $n = 61,$ $\bar{X} = 14.4,$ $s^2 = 2.44;$

(f) $n = 13,$ $\bar{X} = 71.6,$ $\displaystyle\sum_{i=1}^{13}(X_i - \bar{X})^2 = 14.04;$

(g) $n = 10,$ $\bar{X} = 1.1,$ $\displaystyle\sum_{i=1}^{10}X_i^2 = 26.5.$

(2)   Compute 99 percent confidence interval estimates of μ for each of the situations in Exercise (1) and compare the lengths of the intervals with those obtained in answer to Exercise (1).

(3)   A laboratory-type problem: An industrial process has been changed, resulting in a change in the distribution of a certain characteristic of the units produced. From the first dozen units produced by the new process it is desired to estimate, with a 95 percent confidence interval, the mean of the characteristic for the units. Based on the following observations, compute the desired confidence interval. Determine the length of your confidence interval. What assumptions did you make in order to work this problem?

$X_1 = 3.81,$ $X_4 = 4.92,$ $X_7 = 4.52,$ $X_{10} = 6.84,$
$X_2 = 5.07,$ $X_5 = 5.98,$ $X_8 = 4.10,$ $X_{11} = 5.45,$
$X_3 = 6.02,$ $X_6 = 6.35,$ $X_9 = 5.25,$ $X_{12} = 8.00.$

(4)   Assuming that in each of the following situations the observations constitute a random sample from a normal population, test H_0 against H_1 by computing the appropriate test statistic and comparing it with the proper tabulated value.

(a) H_0: $\mu = 5$ versus H_1: $\mu = 7$, $\alpha = .05$, $X_1 = 5.2$, $X_2 = 6.8$, $X_3 = 6.6$;

(b) H_0: $\mu = 6$ versus H_1: $\mu = 10$, $\alpha = .05$, $X_i = \{6.3, 7.9, 8.8, 7.6\}$;

(c) H_0: $\mu = 30$ versus H_1: $\mu = 32$, $\alpha = .05$, $\bar{X} = 31.4$, $n = 40$, $s^2 = 6.4$;

(d) H_0: $\mu = 2$ versus H_1: $\mu = 3$, $\alpha = .01$, $\bar{X} = 2.9$, $n = 30$,
$$\sum_{i=1}^{30} (X_i - \bar{X})^2 = 78.3;$$

(e) H_0: $\mu = 2$ versus H_1: $\mu = 3$, $\alpha = .05$, $\bar{X} = 2.9$, $n = 30$,
$$\sum_{i=1}^{30} (X_i - \bar{X})^2 = 313.2;$$

(f) H_0: $\mu = 19$ versus H_1: $\mu = 18$, $\alpha = .01$, $\bar{X} = 18.1$, $n = 10$,
$$\sum_{i=1}^{10} X^2_i = 3420.1;$$

(g) H_0: $\mu = 19$ versus H_1: $\mu = 18$, $\alpha = .05$, $\bar{X} = 18.2$, $\sigma^2 = 1.225$.

(5)   A laboratory-type problem: Suppose that a teacher feels that she has a class that is better than a typical class. In fact, suppose that the teacher has conjectured that the average IQ for the class is 110. She gives her class of 14 students an IQ test that is devised in such a way that a typical score for the population of students throughout the entire country is 100. Formulate a statistical hypothesis-testing problem. Test the null hypothesis using the following 14 IQ test scores: 109, 117, 129, 108, 94, 120, 103, 131, 107, 115, 101, 98, 97, 112.

2.13   COMPOSITE HYPOTHESES

The theory that has been discussed in connection with simple-versus-simple hypothesis-testing problems can be applied in certain simple-versus-composite situations. Suppose that H_0 is $\mu = \mu_0$, H_1 is $\mu > \mu_0$ and a random observation X is taken from the population under consideration. Suppose that the population is $N(\mu, 1)$. If one particular value, say μ_1, is selected from the set of alternative hypotheses $\mu > \mu_0$, then applying the fundamental principle as discussed in Section 2.9, the null hypothesis H_0 is rejected in favor of μ_1 when $\sqrt{n}\,(\bar{X} - \mu_0)$ is greater than the appropriate tabulated normal value. Provided that the probability of a Type I error is held constant, the critical region remains the same for all values of μ_1, but the probability of a Type II error does depend, of course, on the particular value chosen in the alternative, and because of this we write $\beta(\mu_1)$ for the probability of a Type II error. If the alternative hypothesis is $\mu < \mu_0$, then the points under the left-hand tail of the standard normal constitute the critical region.

When the alternative to H_0 consists of parameter points, some greater than μ_0, the others less than μ_0, the application of the fundamental principle does not help in the selection of a critical region. Faced with an alternative hypothesis such as $\mu \neq \mu_0$, the experimenter usually selects for the critical region values corresponding to both tails of the distribution. This is done in order to provide protection against $\beta(\mu_1)$ being near 1 for certain values of μ_1. Tests of this sort are called two-tailed tests.

The preceding ideas will now be illustrated by letting $\sigma^2 = 1$, $\alpha = .05$, $\mu_0 = 3$, and $n = 1$. Consider the alternative $\mu > 3$ and suppose that the observation is normal. When H_0 is true, $X \sim N(3, 1)$. This density and that corresponding to an arbitrary value of μ_1 appear in Figure 2.5. In Figure 2.6 the probability of a Type II error is shown as a function of

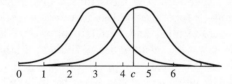

Figure 2.5

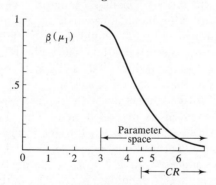

Figure 2.6

μ_1. The construction of the curve $\beta(\mu_1)$ can be facilitated by the application of the normal table. The critical point c is obtained by observing that, when H_0 is true, $\alpha = .05 = P[X > c] = P[X - 3 > c - 3] =$

$P[Z > c - 3]$ where $Z \sim N(0, 1)$. The normal table tells us that $P[Z > 1.645] = .05$ so we conclude that $c = 4.645$. When $X \sim N(\mu_1, 1)$, $\beta(\mu_1) = P[X < c] = P[X - \mu_1 < c - \mu_1] = P[Z < 4.645 - \mu_1]$. The function $\beta(\mu_1)$ can be sketched by selecting values of μ_1 and reading from the normal table the probability corresponding to $4.645 - \mu_1$.

There is no fundamental principle for handling composite-versus-composite testing situations, but there are several indirect ways of attacking the problems. When a simple hypothesis is in some sense typical of the set of hypotheses making up a composite hypothesis, the typical hypothesis is sometimes used to formulate a test procedure. In other cases an experimenter may choose to work with the simple hypothesis in the null hypothesis, which is most difficult to distinguish from the alternative hypothesis. For example, when $H_0: \mu \leq a$ and $H_1: \mu > a$, the hypothesis $\mu = a$ is simple and is the hypothesis most difficult to distinguish from H_1. A test devised for $\mu = a$ versus $\mu > a$ has intitutive appeal for the $\mu \leq a$ versus $\mu > a$ situation.

Let us suppose that the science-test scores for two methods of teaching facts related to weather form the basis for an investigation of the two methods of teaching. Suppose that past test scores associated with the classical method were studied and found to be nicely described by the normal probability law with mean equal to 70 and variance equal to 100. Before the new method is tried, a supporter of the new method suggests that the mean for the new method is greater than 70, while an adherent of the status quo says that the new mean $\mu \leq 70$. Both agree that the change in method will not significantly change the variance, and that in any test procedure or probability calculations calling for a value of the variance, the number $\sigma^2 = 100$ should be used. It is further agreed that the sample size for the new method should be 16, that before the experiment is run a test procedure with a Type II probability curve should be devised, and that α should equal .1.

We now proceed with a discussion of a popular way to handle an experimental problem such as described here. The simple hypothesis $\mu = 70$ is the hypothesis from $H_0: \mu \leq 70$, which is most difficult to distinguish from simple hypotheses among $H_1: \mu > 70$. Using the fundamental principle in connection with $\mu = \mu_0 = 70$ versus $\mu = \mu_1 > 70$, a best test can be devised. If Z denotes the $N(0, 1)$ variable, then reading from Table I, $P[Z > 1.28]$ is approximately equal to $\alpha = .1$. Since $[\sqrt{n}(\bar{X} - 70)]/\sigma \sim N(0, 1)$, rejection of $\mu = 70$ is called for when $\bar{X} > 70 + (1.28)/.4 = 73.2$.

Let us now apply the above rule to the original composite-versus-composite problem. We accept $\mu > 70$ when $\bar{X}$ exceeds 73.2 and accept $\mu \leq 70$ when $\bar{X}$ is less than or equal to 73.2. Now both the Type II error probability and the Type I error probability are functions of the true value of μ. When the true value of μ is ≤ 70, the probability of a Type I error is $\alpha(\mu) = P[\bar{X} > 73.2] = P[.4(\bar{X} - \mu) > .4(73.2 - \mu)] = P[Z > .4(73.2 - \mu)]$ and when the true value of μ is > 70 the probability of a Type II error is $\beta(\mu) = P[Z < .4(73.2 - \mu)]$. Some values of these probabilities are exhibited in Table 2.2.

Table 2.2   The Probability of Errors in a Composite-versus-Composite Situation

True value of μ	Probability of a Type I error $= \alpha(\mu)$	True value of μ	Probability of a Type II error $= \beta(\mu)$
67	.0066	71	.8106
68	.0188	72	.6844
69	.0465	73	.5219
70	.1003	74	.3745
		75	.2358
		76	.1314
		77	.0643
		78	.0274

Table 2.2 reveals that the test rule that we devised keeps the Type I error probability small but allows for large probabilities associated with Type II errors. This could be interpreted as being unfair to the supporter of the new method of teaching. As a matter of fact, in order to be entirely fair, the critical point should be placed at $\mu = 70$. When this is done, both $\alpha(\mu)$ and $\beta(\mu)$ approach $\frac{1}{2}$ as μ approaches 70, but neither of the probabilities become greater than $\frac{1}{2}$. The values of $\alpha(\mu)$ or $\beta(\mu)$, associated with several values of μ, are asked for in the next problem section.

2.14   EXERCISES

(1)   Compute $\alpha(\mu)$ for $\mu = 69, 68$, and 67 when the critical point is chosen to be 70, in the composite-versus-composite illustration of Section 2.13.

(2)   Compute $\beta(\mu)$ for $\mu = 71, 72, 74$, and 77 when the critical point is chosen to be 70, in the composite-versus-composite illustration of Section 2.13.

(3)   Sketch the curve for the probability of a Type II error as a function of $\mu_1 > \mu_0$ when $\mu_0 = 5$, $\alpha = .05$, $\sigma^2 = 4$, and $n = 16$. (Repeat for $n = 25$.)

(4)   Sketch the curve for the probability of a Type II error as a function of $\mu_1 > \mu_0$ when $\mu_0 = 5$, $\alpha = .01$, $\sigma^2 = 4$, and $n = 16$. (Repeat for $n = 25$.)

(5) Set up a two-tailed test for the following composite-versus-composite situation by choosing the midpoint m of the null hypothesis parameter interval as the typical simple null hypothesis. Let $\alpha(\mu) = .05$ when $\mu = m$. Let $n = 9$, $\sigma^2 = 25$, H_0: $5 \leq \mu \leq 7$, and H_1: $\mu > 7$ or $\mu < 5$. Draw a picture of the parameter space, compute $\beta(\mu)$ when $\mu = 3, 4, 5, 7, 8, 9$ and sketch the probability curves $\alpha(\mu)$ and $\beta(\mu)$.

2.15 THE CHI SQUARE DISTRIBUTION
AND SOME APPLICATIONS

Thus far, we have considered one of three important families of distributions derived from a parent normal distribution, that being the Student's t family indexed by the number of degrees of freedom. Our discussion now turns to the *Chi Square family* of distributions. Much earlier, we observed that s^2 was a random variable, indeed a "variable random variable" when the sample size was small. We could study the distribution of s^2 but convention, as well as other reasons, dictates that

$$W = \frac{(n-1)s^2}{\sigma^2}$$

be studied instead. Actually, we study a family of distributions because W, like t, is indexed by the number of degrees of freedom.

To begin our study of the Chi Square family, consider independent random samples of size three taken from an $N(\mu, \sigma^2)$ population. If we compute for many samples the values of the statistic

$$U = \sum_{i=1}^{3} \left(\frac{X_i - \mu}{\sigma}\right)^2,$$

we have in hand specific values of a random variable which can take on only nonnegative numbers. The random variable U has been given the name *Chi Square with three degrees of freedom*. A histogram of the values of U, computed from the many samples of size three, would tend to agree with the asymmetric density curve derived by statisticians and pictured in Figure 2.7. If independent random samples of size five instead of size three are taken, the random variable

$$V = \sum_{i=1}^{5} \left(\frac{X_i - \mu}{\sigma}\right)^2$$

will have the second density curve of Figure 2.7. This member of the family will be identified by the label "five degrees of freedom." In the

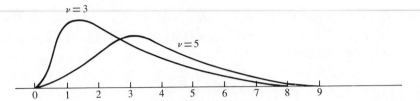

Figure 2.7   Two Chi Square Curves.

exercises to follow, a classroom procedure is presented for empirically obtaining histograms of the Chi Square type.

For each sample size $n = v$, we have a different Chi Square random variable when

$$\sum_{i=1}^{n} \left[\frac{X_i - \mu}{\sigma} \right]^2$$

is formed. We shall denote the family by $\chi^2(v)$ and Table II gives areas under $\chi^2(v)$ curves for many values of v. The $\chi^2(v)$ density function for a specific value of v has a unique maximum at $v - 2$ if v is greater than one, the mean of a $\chi^2(v)$ variable works out to be v and the variance of a $\chi^2(v)$ variable is $2v$. Note that by definition a $\chi^2(v)$ *variable* is the sum of squares of v independent standardized normal random variables.

In some areas of statistical work, the $\chi^2(v)$ family of density curves, or curves similar to these, serve as statistical models with which to describe adequately "real-world" populations of interest. For our purposes, a χ^2 distribution plays the role of a derived distribution, which is studied in order to aid us in making inference statements relative to populations described by the normal probability law. In conjunction with this role, a central concept in much of statistical inference, as we know it today, now needs to be developed.

In the same way that replacement of σ by s in the expression $[(\bar{X} - \mu) \sqrt{n}]/\sigma$ caused us to focus on $[(\bar{X} - \mu) \sqrt{n}]/s$, we are led to consider

$$\sum_{i=1}^{n} \left(\frac{X_i - \bar{X}}{\sigma} \right)^2$$

when we replace μ by $\bar{X}$ in the Chi Square statistic

$$\sum_{i=1}^{n} \left(\frac{X_i - \mu}{\sigma} \right)^2 .$$

We learned earlier that although $[(\bar{X} - \mu)\sqrt{n}]/\sigma$ is normal, $[(\bar{X} - \mu)\sqrt{n}]/s$ is not normal. Reflecting on this fact, it would perhaps seem natural to the reader that

$$\sum_{i=1}^{n}\left(\frac{X_i - \bar{X}}{\sigma}\right)^2$$

should have a distribution somewhat different from that of

$$\sum_{i=1}^{n}\left(\frac{X_i - \mu}{\sigma}\right)^2.$$

The distributions are different, yet amazingly they belong to the same family of distributions. The important fact concerning the distribution of

$$\sum_{i=1}^{n}\left(\frac{X_i - \bar{X}}{\sigma}\right)^2$$

is found in Theorem 2.3.

Theorem 2.3   *If* $X_i \sim NID\ (\mu, \sigma^2)$ *then* $\displaystyle\sum_{i=1}^{n}\left(\frac{X_i - \bar{X}}{\sigma}\right)^2 \sim \chi^2(n - 1).$

Recall that the symbol "$\sim$" is read, "is distributed as." We shall not attempt here to give one of the elegant proofs of this theorem.

As a consequence of the fact that $(n - 1)s^2 = \displaystyle\sum_{i=1}^{n}(X_i - \bar{X})^2$, we have the following corollary to Theorem 2.3:

Corollary $[(n - 1)s^2]/\sigma^2 \sim \chi^2(n - 1)$ *when* s^2 *is computed from a sample* $X_i \sim NID(\mu, \sigma^2)$, $i = 1, 2, \ldots, n$.

The theorem now will be applied to the problem of using a random sample $X_i \sim N(\mu, \sigma^2)$ to set a 100γ percent confidence interval on σ^2. Reading from the line for the appropriate degrees of freedom in the Chi Square table, we can find two numbers l and r such that (see Figure 2.8)

$$P\left[l < \sum_{i=1}^{n}\frac{(X_i - \bar{X})^2}{\sigma^2} < r\right] = \gamma.$$

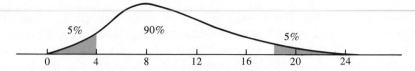

Figure 2.8   The $\chi^2(10)$ Density Curve.

An equivalent probability statement is

$$P\left[\frac{1}{r} < \frac{\sigma^2}{\displaystyle\sum_{i=1}^{n}(X_i - \bar{X})^2} < \frac{1}{l}\right] = \gamma.$$

Finally, we write

$$P\left[\sum_{i=1}^{n}\frac{(X_i - \bar{X})^2}{r} < \sigma^2 < \sum_{i=1}^{n}\frac{(X_i - \bar{X})^2}{l}\right] = \gamma.$$

Observe that the tabulated number l associated with the left tail of the $\chi^2(n-1)$ distribution ends up in the denominator of the upper limit of the interval, and the number r associated with the right tail of the $\chi^2(n-1)$ distribution finds itself in the denominator of the lower limit of the interval. The random length of intervals constructed in this way is

$$L = \sum_{i=1}^{n}(X_i - \bar{X})^2\left(\frac{1}{l} - \frac{1}{r}\right).$$

The expected or average length can be decreased by decreasing γ and increasing ν, for such changes in γ and ν decrease the quantity $[1/l - 1/r]$ faster than the expected value of $\sum_{i=1}^{n}(X_i - \bar{X})^2$ is increased.

An illustration of the described procedure will now be presented. Suppose that a random sample of size 11 yields $\sum_{i=1}^{11}(X_i - \bar{X})^2 = 36.62$. If γ is chosen to be .90 then, reading from the Chi Square table with $\nu = 10$, we obtain $l = 3.94$ and $r = 18.31$, as shown in Figure 2.8. The calculated confidence statement is $P[2 < \sigma^2 < 9.3] = .90$ with length $L = 7.3$. Notice that $s^2 = 3.662$ is not at the midpoint of this interval.

A slightly shorter interval could have been obtained by assigning the probability $(1 - \gamma)$ in the two tails, in such a way that $[1/l - 1/r]$ is minimized; but this is seldom done because, first, appropriate tables are not available and, second, the exact amount to go into each tail depends on γ and ν and cannot be computed readily. Except when ν is very small, partitioning $(1 - \gamma)$ equally between the two tails gives intervals with expected length near the minimum expected length.

The situation with regard to testing hypotheses concerning σ^2, from information in a random sample of size n from $N(\mu, \sigma^2)$, may be summarized as follows. When μ is known and $\sigma^2{}_0$ is the hypothesized value of σ^2, the appropriate test statistic is

$$\sum_{i=1}^{n} \left(\frac{X_i - \mu}{\sigma_0} \right)^2$$

and the proper tabulated χ^2 value has n degrees of freedom. When μ is unknown, the appropriate test statistic is

$$\sum_{i=1}^{n} \left(\frac{X_i - \bar{X}}{\sigma_0} \right)^2$$

and the proper tabulated χ^2 value has $(n - 1)$ degrees of freedom. In either case, the nature of the critical region in the sample space of the test statistic depends on the form of the alternative hypothesis. If H_1 is $\sigma^2 \neq \sigma^2{}_0$, a two-tailed test is called for. If H_1 is $\sigma^2 > \sigma^2{}_0$, rejection of H_0 follows only for values of the test statistic in the upper tail, and if H_1 is $\sigma^2 < \sigma^2{}_0$, rejection of H_0 is called for only when values of the test statistic fall in the lower tail.

To illustrate the procedure, consider a sample of size $n = 21$ from a $N(\mu, \sigma^2)$ population with μ unknown. Suppose that the sample yield $\sum_{i=1}^{21} (X_i - \bar{X})^2 = 288.5$ and it was desired to test $H_0: \sigma^2 = 25$ against $H_1: \sigma^2 < 25$. Letting $\alpha = .05$, the tabulated $\chi^2(20)$ value is 10.85. The nature of the alternative hypothesis demands rejection if the test statistic has a value less than 10.85. An unbiased estimate of σ^2 is the value $s^2 = 14.425$. This appears considerably different from the hypothesized value $\sigma^2{}_0 = 25$, yet the value of the test statistic $\sum_{i=1}^{21} [(X_i - \bar{X})/\sigma_0]^2$ is 11.54 which does not fall in the critical region. Our conclusion must then be that the difference between s^2 and σ_0^2 is not significant. We have not gathered, in our sample, sufficient evidence to support the rejection of H_0.

2.16 EXERCISES INVOLVING INFERENCE RELATIVE TO THE VARIANCE

In the following problems, consider the observations $X_i \sim \text{NID}(\mu, \sigma^2)$.

(1) A laboratory-type exercise: Draw random samples of size three from the "normal" population of Table V. Compute for each sample the value of

$$U = \sum_{i=1}^{3} [(X_i - \mu)/\sigma]^2$$

and tabulate the values for the class. Denote by an area of $1/k$ units each value of U, thereby creating a histogram. Compare the histogram with the Chi Square curve pictured in Figure 2.7. Compute the sample mean $\bar{U}$ and the sample variance s^2_U and compare these with the population mean $\nu = 3$ and the population variance $2\nu = 6$.

(2) Determine a 95 percent confidence interval estimate for σ^2 if $X_1 = 4$, $X_2 = 2$, $X_3 = 1$, $X_4 = 6$, and $X_5 = 2$.

(3) Compute a 99 percent interval estimate of σ^2 when $\sum_{i=1}^{10} X_i = 6.5$ and $\sum_{i=1}^{10} X^2_i = 9.225$.

(4) Compute 94 percent interval estimates of σ^2 when $\sum_{i=1}^{6} X_i = 12$ and $\sum_{i=1}^{6} X^2_i = 26$ by assigning in turn (a) 1 percent, (b) 5 percent, (c) 6 percent, (d) 3.5 percent, and (e) 2.5 percent of the probability to the left tail and the rest to the right tail.

(5) With $\alpha = .1$, test $\sigma^2 = 40$ against $\sigma^2 \neq 40$ when the observations are 2, 7, 5, 9, and 7.

(6) With $\alpha = .1$, test $\sigma^2 = 1$ versus $\sigma^2 > 1$ when $\sum_{i=1}^{20} (X_i - \bar{X})^2 = 38$.

(7) With $\alpha = .025$, test $\sigma^2 = 10$ versus $\sigma^2 < 10$ when $\sum_{i=1}^{25} X^2_i = 136$ and $\sum_{i=1}^{25} X_i = 50$.

(8) A laboratory-type problem: An author of examinations claims that a certain examination was devised in such a way that for a randomly selected

group of students, the scores will follow a standardized normal distribution. Scores for a class of 20 students were as follows: .05, .79, −1.12, −.13, 2.07, 3.52, 1.44, −.72, 2.59, 3.08, .55, −.09, −.42, 1.36, 2.15, .98, −.03, .85, .64, and 1.80. Is there reason to believe that either the test was not what it claimed to be or that the class was not a randomly selected group? Support your answer with some statistical procedures.

2.17   POINT ESTIMATION

The study of classical statistical inference can be partitioned rather naturally into three study areas: testing statistical hypotheses, interval estimation, and point estimation. With reference to sampling from one normal population, we have devoted some effort to each of these areas. The point-estimation theory will now be pursued a little further.

In addition to the term "mean" there are, in common use, several other terms meant to convey essentially the same concept as that associated with the word mean. In Chapter 1, mention was made of the equivalence of "mean of a population" and "expected value of a random variable." The concept corresponds to the idea of center of gravity which is studied in physics. Consider, as pictured in Figure 2.9, a horizontal "weightless"

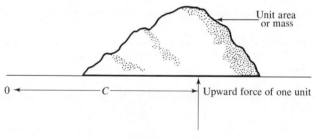

Figure 2.9

line, positioned so as to be free to pivot about the origin. Suppose that a unit of mass is distributed continuously but not necessarily uniformly along the line. An upward force will be required to keep the line in a horizontal position. Suppose that the upward force is restricted to be equal to one unit. The point at which this force should be applied in order to keep equilibrium is called the center of gravity. If the area under the density curve, for a random variable X, is thought of as the mass placed on the line, then the distance from the center of gravity to the origin corresponds to the expected value of X.

Note that if the distribution of X is symmetric, then it is symmetric about μ. The Student t variable is of this type with symmetry about zero.

A frequently used notation for the expression "the expected value of X" is $\mathcal{E}(X)$. In this notation we can neatly express some ideas that have been discussed several times.

Statement: *For any set of random variables $X_1, \ldots, X_n$, with means $\mu_1, \mu_2, \ldots, \mu_n$, and any real constants $c_1, c_2, \ldots, c_n$,*

$$\mathcal{E}\left[\sum_{i=1}^{k} c_i X_i \right] = \sum_{i=1}^{k} c_i \mathcal{E}(X_i) = \sum_{i=1}^{k} c_i \mu_i.$$

Some special cases are:

$$\mathcal{E}[X_1 + X_2] = \mathcal{E}[X_1] + \mathcal{E}[X_2];$$

$$\mathcal{E}[cX] = c\mathcal{E}(X);$$

if $\mathcal{E}[X_i] = \mu$ for each $i = 1, 2, \ldots, n$, then $\mathcal{E} \sum_{i=1}^{n} X_i = n\mu$; and if

$\mathcal{E}[X_i] = \mu$ for each $i = 1, 2, \ldots, n$ and $\bar{X} = \dfrac{1}{n} \sum_{i=1}^{n} X_i$, then $\mathcal{E}[\bar{X}] = \mu$.

When $\bar{X}$ is used as an estimator of μ, the property expressed in the last special case above is often described by saying that $\bar{X}$ is an *unbiased* estimator of μ. Consider a more general situation where $Y_1, Y_2, \ldots, Y_n$ are independent observational random variables from a population whose density involves an unknown parameter denoted by θ. Let $\hat{\theta}$ be some function of the random variables $Y_1, Y_2, \ldots, Y_n$. If $\mathcal{E}(\hat{\theta}) = \theta$, then $\hat{\theta}$ is said to be an *unbiased* estimator of θ, but if $\mathcal{E}(\hat{\theta}) \neq \theta$ then $\hat{\theta}$ is said to be *biased* and the *bias* $= \theta - \mathcal{E}(\hat{\theta})$.

When we sample at random from a normal population,

$$S^2 = \frac{1}{n-1} \sum_{i=1}^{n} (X_i - \bar{X})^2$$

is an unbiased estimator of σ^2. This can be argued, using elementary methods, if we start with the fact that, when $W \sim \chi^2(\nu)$ then $\mathcal{E}[W] = \nu$. Since

$$\frac{(n-1)S^2}{\sigma^2} \sim \chi^2(n-1)$$

then

$$\mathcal{E}\left[\frac{(n-1)S^2}{\sigma^2}\right] = n - 1, \quad \mathcal{E}\left[\frac{S^2}{\sigma^2}\right] = 1 \quad \text{and} \quad \mathcal{E}[S^2] = \sigma^2.$$

The last equation asserts that S^2 is an unbiased estimator of σ^2. This explains, at least in part, the reason for defining S^2 with $(n-1)$ as the divisor instead of n. The variable $\sum_{i=1}^{n} (X_i - \bar{X})^2/n$, when used as an estimator of σ^2 has a bias equal to σ^2/n.

Although S^2 is an unbiased estimator of σ^2, the standard deviation S is a biased estimator of σ. The bias in S depends on the sample size and decreases (approaches zero) as n grows large. The correction that can be made to remove the bias is not a simple function of n and will not be presented.

Throughout this chapter, attention has been focused on sampling from one population and much of the discussion has centered around the random variable $\bar{X}$. $\bar{X}$ can be thought of as a very special *linear combination* of $X_1, X_2, \ldots, X_n$, a *linear combination* being a function of the form

$$\sum_{i=1}^{n} c_i X_i.$$

When $\bar{X}$ is thought of as a linear combination, each c_i becomes $1/n$ and we write

$$\bar{X} = \frac{1}{n} X_1 + \frac{1}{n} X_2 + \cdots + \frac{1}{n} X_n.$$

The constants c_i, $i = 1, 2, \ldots, n$ are referred to as weights.

It is of course natural, in a symmetric situation, to weight each observation equally. Later in the text we shall encounter situations where observations quite naturally should and do carry different weights. An example of a linear combination with unequal weights is

$$Y = \frac{1}{5} X_1 + \frac{2}{5} X_2 + \frac{2}{5} X_3.$$

This linear combination and that of $\bar{X}$ both have the property that

$$\sum_{i=1}^{n} c_i = 1.$$

The general problem of finding the variance of a linear combination in terms of the variance of the individual random variables will not be discussed in this chapter, but a special case of their relationship appears in Theorem 2.4. The statement of the theorem serves as a very important computational rule.

Theorem 2.4 *Let c_i, $i = 1, 2, \ldots, n$ be any set of real numbers. For any random sample $X_1, X_2, \ldots, X_n$ taken from a population with variance σ^2*

$$Var\left[\sum_{i=1}^{n} c_i X_i\right] = \sum_{i=1}^{n} c^2_i \, Var\,(X_i) = \sigma^2 \sum_{i=1}^{n} c^2_i.$$

This theorem will not be proved but it can be made more intuitively appealing by considering the following facts: First, the rule when applied to $\bar{X}$ gives

$$Var\,[\bar{X}] = Var\left[\sum_{i=1}^{n} \frac{1}{n} X_i\right] = \frac{1}{n^2} \sum_{i=1}^{n} Var\,(X_i) = \frac{\sigma^2}{n}$$

and second, since we have seen that $S^2_{cX} = c^2 S^2_X$ it seems justifiable to have $Var\,(cX) = c^2\,Var\,(X)$.

When sampling at random from one population, another of the good properties enjoyed by $\hat{\mu} = \bar{X}$ is that it is the unbiased estimator of the form $\sum_{i=1}^{n} c_i X_i$ possessing *minimum variance*. In order to argue this fact, note that unbiased implies $\sum_{i=1}^{n} c_i = 1$. The variance of $\sum_{i=1}^{n} c_i X_i = \sigma^2 \sum_{i=1}^{n} c^2_i$. We desire to minimize $\sum_{i=1}^{n} c^2_i$ subject to the restriction that $\sum_{i=1}^{n} c_i = 1$. Setting $c_i = 1/n$ for each i satisfies the restriction and corresponds to a variance of σ^2/n. Let $c_i - (1/n) = d_i$. We then have $\sum_{i=1}^{n} d_i = 0$ and

$$\sum_{i=1}^{n} c^2_i = \sum_{i=1}^{n} \left(d_i + \frac{1}{n}\right)^2 = \sum_{i=1}^{n} d^2_i + \frac{1}{n}.$$

Thus $\sum_{i=1}^{n} c^2_i \geq 1/n$ for any set of c_i such that $\sum_{i=1}^{n} c_i = 1$, and $\bar{X}$ has minimum variance.

The mean for a population was not defined in Chapter 1. It was referred to as the population counterpart of the sample mean. The fact that the variance of $\bar{X}$ approaches zero as $n \to \infty$ can be used to give a definition of μ. μ is that value which $\bar{X}$ approaches as $n \to \infty$. (We are of course considering one infinite population and $\bar{X}$ is computed from a random sample.)

The variance of a random variable X can be conveniently expressed in expected value notation. The relationship is

$$\sigma_X{}^2 = \mathcal{E}[X - \mathcal{E}(X)]^2.$$

This equation in effect states that the variance of X is the same as the mean of the random variable $Y = [X - \mathcal{E}(X)]^2$. The details for Exercise (10) of Section 2.18 should provide some insights into why this relationship is true in the normal case. Also left as an exercise is the problem of showing that

$$\sigma_X{}^2 = \mathcal{E}(X^2) - [\mathcal{E}(X)]^2.$$

Since $X - \mathcal{E}(X)$ is a deviation about the mean of X, $\sigma^2{}_X$ is often called the second moment about the mean. Of course, it is then quite natural to call the number $\mathcal{E}[X - \theta]^2$ the second moment about θ. Consider again the situation where the random variable $\hat{\theta}$ is a function of n observational random variables $Y_1, Y_2, \ldots, Y_n$ each taken from a population with parameter θ. The number $\mathcal{E}[\hat{\theta} - \theta]^2$ is called the *mean squared error of* $\hat{\theta}$. We shall use the abbreviation MSE. In the case where $\mathcal{E}(\hat{\theta}) = \theta$, the MSE of $\hat{\theta}$ coincides with the variance of $\hat{\theta}$, but when $\hat{\theta}$ is a biased estimator of θ, the MSE is greater than the variance. The interesting and important relationship among MSE, variance, and bias is expressed in Theorem 2.5.

Theorem 2.5 *Let $\hat{\theta}$ be an estimator of θ, then*

$$MSE \text{ of } \hat{\theta} = Var \ (\hat{\theta}) + (Bias)^2.$$

To prove this theorem we write

$$\text{MSE} = \mathcal{E}[\hat{\theta} - \theta]^2 = \mathcal{E}\{[\hat{\theta} - \mathcal{E}(\hat{\theta})] - [\theta - \mathcal{E}(\hat{\theta})]\}^2.$$

Then

$$\text{MSE} = \mathcal{E}\{[\hat{\theta} - \mathcal{E}(\hat{\theta})]^2 - 2[\hat{\theta} - \mathcal{E}(\hat{\theta})][\theta - \mathcal{E}(\hat{\theta})] + [\theta - \mathcal{E}(\hat{\theta})]^2\}.$$

$$\text{MSE} = \mathcal{E}[\hat{\theta} - \mathcal{E}(\hat{\theta})]^2 - 2[\theta - \mathcal{E}(\hat{\theta})][\hat{\theta} - \mathcal{E}(\hat{\theta})] + \mathcal{E}[\theta - \mathcal{E}(\hat{\theta})]^2.$$

$$\text{MSE} = Var \ (\hat{\theta}) - 2 \text{ Bias } [\mathcal{E}(\hat{\theta}) - \mathcal{E}(\hat{\theta})] + [\theta - \mathcal{E}(\hat{\theta})]^2$$

and finally we have MSE $= Var \ (\hat{\theta}) + (Bias)^2$.

The concept here can be illustrated by creating a picture typical of the field layout in army artillery fire. Let T denote the target location, let A denote the aiming point, and suppose that six shells fired from a weapon had impact points as indicated (Figure 2.10). Focus attention on one random impact point as indicated in Figure 2.10. The length b of the line from T to A is the bias in the weapon and aiming apparatus. The random distance from T to an impact point I is the error made on that particular

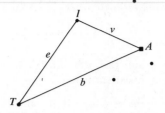

Figure 2.10

firing of the weapon. The expected value of the squared variable distances from A to impact point I is a quantity analogous to the concept of variance.

2.18 EXERCISES

(1) Suppose that $X_i \sim \mathrm{NID}(\mu, \sigma^2)$ for $i = 1, 2, 3$. Determine the expected value and the variance for each of the following linear functions of the X_i.
   (a) $\frac{1}{2}X_1 + \frac{1}{4}X_2 + \frac{1}{4}X_3$
   (b) $\frac{1}{9}X_1 + \frac{1}{9}X_2 + (\frac{7}{9})X_3$
   (c) $\frac{1}{3}X_1 + \frac{1}{3}X_2 + \frac{1}{3}X_3$
   (d) $\frac{1}{2}X_1 + \frac{1}{2}X_2$

(2) Use an expected value argument to establish that $\mathcal{E}(Z) = 0$, and Var $(Z) = 1$ when $Z = (X - \mu)/\sigma$ and $X \sim N(\mu, \sigma^2)$.

(3) The median of a population is a number m such that $P[X < m] = \frac{1}{2}$. Comment on the fact that when the population is symmetric and $\bar{X}$ is computed from a random sample, $\bar{X}$ is an unbiased estimator of m.

(4) Let X_1 be a random observation from a particular population with mean μ. Consider the problem of estimating μ when the second observation has expectation $\mu + b$, and in general $\mathcal{E}[X_k] = \mu + (k - 1)b$. In other words, consider independent observations with homogeneous variance but with the property that each observation after the first one is more biased than the previous observation. $\bar{X} = (1/n) \sum_{k=1}^{n} X_k$ would of course be a biased estimator of μ. Determine this bias and determine the MSE of $\bar{X}$ as an estimator of μ.

(5) Suppose that a random sample of size two was taken from a normal population with mean μ but somehow X_2, the larger of the observations, was destroyed or lost. Can you argue that the smaller remaining observation X_1 is now a biased estimator of μ?

(6) Let k and a be two fixed real numbers. Suppose that an experimenter

wishes to base his unbiased estimate of $(k\mu + a)$ on a random sample of size n from a population $N(\mu, \sigma^2)$. Argue that $a + (k/n) \sum\limits_{i=1}^{n} X_i$ is an unbiased estimate of $k\mu + a$ and that this statistic has variance less than that of any other unbiased linear statistic.

(7)   Suppose $W_1 \sim \chi^2(k_1)$ and $W_2 \sim \chi^2(k_2)$. If W_1 and W_2 are independent, their sum $(W_1 + W_2)$ is a Chi Square variable with k degrees of freedom. Use an expected value argument to establish that $k = k_1 + k_2$.

(8)   Let $X_1, X_2, \ldots, X_n$ be a random sample from an infinite population. Argue that

$$\frac{1}{n(n-1)} \sum_{i=1}^{n} (X_i - \bar{X})^2$$

is an unbiased estimator of the variance of $\bar{X}$.

(9)   Use the fact that S^2 is an unbiased estimator of σ^2 to show that when $(1/n) \sum\limits_{i=1}^{n} (X_i - \bar{X})^2$ is used to estimate σ^2, the bias equals σ^2/n.

(10)   Use the fact that $W \sim \chi^2(n)$ implies $\mathcal{E}(W) = n$, to prove that for a normal variable X, $\sigma^2{}_X = \mathcal{E}[X - \mathcal{E}(X)]^2$.

(11)   Starting at $\sigma^2{}_X = \mathcal{E}[X - \mathcal{E}(X)]^2$, show that $\sigma^2{}_X = \mathcal{E}(X^2) - (\mathcal{E}X)^2$.

(12)   Use the fact that the variance of a Chi Square variable is twice its degrees of freedom to determine the variance of S^2, when S^2 is a function of normal and independent random variables.

(13)   Use expected value notation to determine $\mathcal{E}[S^4]$ when S^2 is computed from random normal data.

(14)   Let $X_i \sim \text{NID}(\mu, \sigma^2)$. Determine the MSE of $(1/n) \sum\limits_{i=1}^{n} (X_i - \bar{X})^2$ when used as an estimator of σ^2.

(15)   Suppose that S^2 and $(1/n) \sum\limits_{i=1}^{n} (X_i - \bar{X})^2$ are computed from normal data and both are used as estimates of σ^2. Compare their MSE's for sample sizes $n = 2, 4$, and 10.

(16)   Suppose that it is known from theoretical considerations that a population is symmetric about zero. Suppose too that this population is adequately described by a member of the normal family of densities. Let $X_1, \ldots, X_n$ be a random sample from this population. Use the fact that $X^2{}_i/\sigma^2 \sim$

$\chi^2(1)$ to prove that $(1/n) \sum\limits_{i=1}^{n} X^2{}_i$ is an unbiased estimator of σ^2.

(17)   For the situation in Exercise (16) determine the Var $[(1/n)\Sigma X^2{}_i]$ and compare it with the Var (S^2).

(18)   Consider the problem of estimating μ and σ^2 in a normal population when, because of a faulty measuring device, the random observations X_1, X_2, . . . , X_n are each recorded an unknown but fixed c units greater than they should have been. Show that $\bar{X}$ is a biased estimator of μ but that S^2 remains an unbiased estimator of σ^2.

(19)   Suppose that for the situation described in exercise (18) each observation is recorded to be d times what it should have been. Determine the bias in $\bar{X}$ and S^2 as estimators of μ and σ^2, respectively.

3

STATISTICAL INFERENCE RELATIVE TO SAMPLING FROM TWO NORMAL POPULATIONS

3.1 INTRODUCTION TO THE STUDY OF TWO POPULATIONS

In this chapter attention ultimately will be directed toward applying techniques for testing hypotheses, setting confidence intervals, and determining point estimators in situations where observations come from two normal populations. In striving toward this goal, we first need to prepare ourselves by considering new concepts and making ourselves aware of some additional facts relative to variances and means. To begin with, we shall consider a rather basic theorem in expectation theory.

Theorem 3.1 *Consider independent random variables X and Y. If $U = X + Y$ and $V = X - Y$ then $\sigma^2_U = \sigma^2_X + \sigma^2_Y = \sigma^2_V$.*

The mathematical statistician proves by his knowledge of mathematics that the theorem is true and thereby has confidence in using the formula. The nonmathematician can use a device such as actually drawing samples to give him confidence in the statement, because he can show that a similar relationship holds for the sample variances. Suppose then that we consider a random sample $X_1, \ldots, X_n$ from a population with variance σ^2_X and a random sample $Y_1, \ldots, Y_n$ independent of the first sample but taken from a population with variance σ^2_Y. The sets of observations corresponding to the random variables U and V are then created by defining $U_i = X_i + Y_i$ and $V_i = X_i - Y_i$. Here will be given the details for showing that $s^2_V \doteq s^2_X + s^2_Y$ and the proof that $s^2_U \doteq s^2_X + s^2_Y$ will be left as an exercise. (The symbol $\doteq$ will denote "approximately equal to.") By definition,

$$s^2_V = \frac{1}{n-1} \sum_{i=1}^{n} (V_i - \bar{V})^2.$$

Substituting for V_i and $\bar{V}$, we have

$$s^2_V = \frac{1}{n-1} \sum_{i=1}^{n} (X_i - Y_i - \bar{X} + \bar{Y})^2 = \frac{1}{n-1} \sum_{i=1}^{n} [(X_i - \bar{X}) - (Y_i - \bar{Y})]^2.$$

Expanding the indicated binomial, we have

$$s^2_V = \frac{1}{n-1} \sum_{i=1}^{n} (X_i - \bar{X})^2 + \frac{1}{n-1} \sum_{i=1}^{n} (Y_i - \bar{Y})^2 - \frac{2}{n-1} \sum_{i=1}^{n} (X_i - \bar{X})(Y_i - \bar{Y})$$

or

$$s^2_V = s^2_X + s^2_Y - \frac{2}{n-1} \sum_{i=1}^{n} (X_i - \bar{X})(Y_i - \bar{Y}).$$

It could be verified by repeated sampling from independent populations that the statistic $s_{XY} = \frac{1}{n-1} \sum_{i=1}^{n} (X_i - \bar{X})(Y_i - \bar{Y})$ takes on small values relative to s^2_X and s^2_Y. The statistic, s_{XY}, called the *sample covariance*, takes on other than small values when we concern ourselves with dependent random variables. It plays an important role in later considerations.

Another fact that is used heavily, when attention is turned to applications, is related in the next theorem.

Theorem 3.2   *Let a and b be arbitrary real numbers. If X and Y are normal random variables with means μ_X and μ_Y then $aX + bY$ is a normal random variable with mean $a\mu_X + b\mu_Y$.*

Confidence that the new random variable is normal can be gained for the nonmathematician by repeated sampling, and confidence in the new mean value can further be strengthened by proving that a corresponding relationship holds for the sample mean of observations $aX_i + bY_i$.

3.2 WEIGHTING ESTIMATES OF μ AND δ^2

Suppose that, prior to the final week of the season, a baseball player's batting average was .300 and during the last week of the season his average was .400. His season's average cannot be computed from these facts alone, but it can be said that his season average is between .300 and .400. If it is known that the .300 average was based on $n_1 = 450$ official times at bat and the .400 average was based on $n_2 = 35$ official times at bat, then his seasons average can be computed. The method of obtaining the overall average is the object of our attention. Let h_1 be the number of hits prior to the last week and let h_2 be the number of hits during the last week. His season's average, denoted by a, is then

$$a = \frac{h_1 + h_2}{n_1 + n_2}.$$

Now $h_1/n_1 = .3$ and $h_2/n_2 = .4$ hence $h_1 = .3n_1$ and $h_2 = .4n_2$. Thus

$$a = \frac{.3n_1 + .4n_2}{n_1 + n_2}$$

or

$$a = \frac{n_1}{n_1 + n_2}(.300) + \frac{n_2}{n_1 + n_2}(.400).$$

In the example

$$\frac{n_1}{n_1 + n_2} = \frac{450}{485}$$

and

$$\frac{n_2}{n_1 + n_2} = \frac{35}{485}$$

so that

$$a = \frac{450(.3) + 35(.4)}{485} = \frac{149}{485} = .3072.$$

In retrospect, it can be said that the .300 average carried more weight than did the .400 average.

Consider now the situation where two estimators $\bar{X}_1$ and $\bar{X}_2$ exist for one parameter μ. If $\bar{X}_1$ is based on n_1 observations $\bar{X}_2$ is based on n_2

observations and all observations come from the same population, then the *combined unbiased estimator* of μ with the smallest variance is

$$\frac{n_1}{n_1 + n_2} \bar{X}_1 + \frac{n_2}{n_1 + n_2} \bar{X}_2.$$

Using this estimator is equivalent to finding the grand average of the $n_1 + n_2$ observations.

In some applications, the case where samples come from different populations must be considered. When the populations have a common unknown mean μ but known variances σ^2_1 and σ^2_2, sample information from each population can be combined to obtain an unbiased estimate of μ. The linear estimator with minimum variance is the one where individual means are weighted inversely proportional to their variance. The combined estimator referred to in this case can be written in the form

$$\frac{n_1 \sigma^2_2 \bar{X}_1 + n_2 \sigma^2_1 \bar{X}_2}{n_2 \sigma^2_1 + n_1 \sigma^2_2}.$$

The important question of how to combine estimates of a common mean when the variances are unknown will not be discussed in this text.

In real-world problems, the assumption that two populations have common variance is often justified. When this assumption is made and the common variance is to be estimated from two samples, we are faced again with the problem of how to weight the individual estimates. In many applications, the preferred estimator is

$$s^2_p = \frac{(n_1 - 1)s^2_1 + (n_2 - 1)s^2_2}{n_1 + n_2 - 2} = \frac{\sum\limits_{i=1}^{n_1} (X_{1i} - \bar{X}_1)^2 + \sum\limits_{i=1}^{n_2} (X_{2i} - \bar{X}_2)^2}{n_1 - 1 + n_2 - 1}$$

where X_{1i} is the ith observation in the first sample and X_{2i} is the ith observation in the second sample. S^2_p is called the *pooled estimator* of the common variance and the formula is referred to as the pooled sum of squares divided by the *pooled degrees of freedom*.

3.3 EXERCISES

(1) Let X and Y be independent normal random variables each with mean μ and variance σ^2. State the distribution of

(a) $X + Y$ (b) $X - Y$

(c) $2X + 3Y$ (d) $2X - Y$

(e) $\dfrac{X + Y}{2}$ (f) $\dfrac{X + Y}{2\sigma}$

(2) Let X_i and Y_i be independent random variables with $X_i \sim N(\mu,\ \sigma^2)$ $i = 1, 2, \ldots, 5$ and $Y_i \sim N(\mu, \sigma^2)$ $i = 1, 2, \ldots, 5$. Let $\bar{X} = \frac{1}{5} \sum_{i=1}^{5} X_i$ and $\bar{Y} = \frac{1}{5} \sum_{i=1}^{5} Y_i$. State the distribution of

(a) $\bar{X} - \bar{Y}$ (b) $\bar{X} + \bar{Y}$

(c) $\dfrac{\bar{X} - \bar{Y}}{\sigma}$ (d) $\left(\dfrac{\bar{X} - \bar{Y}}{\sigma}\right) \sqrt{5/2}$

(3) If $X_i \sim \text{NID}(10, 24)i = 1, 2, \ldots, 8$, $Y_i \sim \text{NID}(15, 60)i = 1, 2, \ldots, 5$, and all X_i are independent of the Y_i's, write the distribution of

(a) $X_1 + Y_1$ (b) $X_1 + X_2$

(c) $X_1 - X_2$ (d) $X_1 - Y_1$

(e) $X_1 + X_2 + X_3$ (f) $\sum_{i=1}^{8} X_i$

(g) $\bar{X} = \frac{1}{8} \sum_{i=1}^{8} X_i$ (h) $\bar{Y} = \frac{1}{5} \sum_{i=1}^{5} Y_i$

(i) $\bar{Y} - \bar{X}$ (j) $\bar{Y} + \bar{X}$

(k) $X_1 + X_2 - Y_1 - Y_2$

(4) If $X_i \sim \text{NID}(\mu, \sigma^2)i = 1, 2, \ldots, 10$, $Y_i \sim \text{NID}(\mu, \sigma^2)i = 1, 2, \ldots, 20$ and all X_i are independent of the Y_i's write the distribution of (a) $\bar{X}$, (b) $\bar{Y}$, (c) $\bar{X} - \bar{Y}$, (d) $\dfrac{\bar{X} - \bar{Y}}{\sigma}$, and (e) $\left(\dfrac{\bar{X} - \bar{Y}}{\sigma}\right) \sqrt{\frac{3}{20}}$.

(5) Let $W_i = 2X_i + 3Y_i$, $i = 1, 2, \ldots, 7$. Prove that $\bar{W} = 2\bar{X} + 3\bar{Y}$.

(6) Let $W_i = aX_i + bY_i$, $i = 1, 2, \ldots, n$. Prove that $\bar{W} = a\bar{X} + b\bar{Y}$.

(7) A laboratory-type problem: Let $V_i = X_i - Y_i$ for each of the enumerated observation pairs. From the data compute the sample variances $s^2{}_X$, $s^2{}_Y$, and $s^2{}_V$ and observe the difference between $(s^2{}_X + s^2{}_Y)$ and $s^2{}_V$.

X_i: 6.0, 2.1, 5.1, 9.4, 8.3, 5.6, 4.9, 3.2, 9.8, 7.3, 8.0, 4.1, 5.9, 8.7, 5.6, 4.9.
Y_i: 2.2, 4.0, 8.3, 7.2, 4.2, 3.3, 2.5, 9.7, 6.2, 7.4, 3.2, 5.0, 6.1, 2.1, 5.0, 6.3.

(8) Show that, provided s_{XY} is small, $s^2{}_U$ is approximately equal to $(s^2{}_X + s^2{}_Y)$ where U is the random variable $U = X + Y$.

(9) A football quarterback completed 40 percent of his passes in his first game and 60 percent of his passes in his second game. If he attempted 30 passes in his first game and 15 passes in his second game, what is his overall passing percentage?

(10) Suppose that two samples were taken from the same population. The first

sample, of size 12, yielded an average of $\bar{X}_1 = 3.82$ and the second sample, of size 18, yielded an average of $\bar{X}_2 = 3.54$. Compute the combined unbiased minimum variance estimate of the common mean μ.

(11)   Information concerning a mean μ, common to two populations, comes from a sample of size $n_1 = 24$ from the first population where the variance is $\sigma^2{}_1 = 3$ and from a sample of size $n_2 = 16$ from the second population where the variance is $\sigma^2{}_2 = 8$. If $\bar{X}_1$ is 10 and $\bar{X}_2$ is 20, compute the value of the combined, linear, unbiased, minimum variance estimate of μ.

(12)   A laboratory-type problem: It is known that a certain characteristic of an organ in a particular breed of fish has the properties that, in a relatively small region of the ocean, the mean value μ for males is very nearly equal to the mean value for females but the variance σ^2 for males is about twice the variance for females. After much work, 28 fish were caught and the organ characteristic measured for each fish. With the observations for males denoted by Y and those for females denoted by X, the data was as follows:

Y: 2.70, 2.56, 2.20, 2.93, 2.01, 2.73, 2.62, 2.40, 2.80, 2.99.
X: 2.31, 2.57, 2.41, 2.03, 2.42, 2.51, 2.30, 2.60, 2.21, 2.35, 2.32, 2.42, 2.50, 2.45, 2.10, 2.51, 2.59, 2.39.

Compute a "good" combined estimate of μ.

(13)   Let $X_1 = 7, X_2 = 9, Y_1 = 1, Y_2 = 4, Y_3 = 2$, and suppose $X_i \sim \text{NID}(\mu_1, \sigma^2)$ and $Y_i \sim \text{NID}(\mu_2, \sigma^2)$. From the data, compute $s^2{}_p$, the pooled estimate of σ^2.

(14)   A laboratory-type problem: An economist has reason to believe that the variance of income is about constant in certain northwestern farm towns. He knows too that the mean incomes may differ greatly from town to town. Compute the pooled estimate of the income variance from the following income samples for two farm towns. The data is in thousands of dollars per year, per family.

First town: 10.4, 9.8, 14.2, 16.1, 8.2, 7.6, 14.8, 12.0, 11.2, 8.3, 9.4.
Second town: 7.3, 8.4, 6.8, 5.4, 10.2, 8.7, 4.5, 6.2, 5.8.

(15)   If $s^2{}_1 = 3.2$ and $s^2{}_2 = 3.8$ are both unbiased estimates of σ^2 based on $n_1 = 11$ and $n_2 = 21$ observations, respectively, compute the pooled estimate of σ^2.

(16)   If σ^2 is the common variance of the random variables X and Y, compute the pooled estimate of σ^2 from the following data.

$$\sum_{i=1}^{8} X^2{}_i = 304, \quad \sum_{i=1}^{8} X_i = 48, \quad \sum_{i=1}^{14} Y^2{}_i = 1432, \quad \sum_{i=1}^{14} Y_i = 140.$$

3.4   INFERENCE RELATIVE TO THE MEANS WHEN INDEPENDENT SAMPLES ARE TAKEN AND EQUALITY OF VARIANCE IS ASSUMED

Consider the situation where observations $X_1, X_2, \ldots, X_{n_1}$ come from $N(\mu_X, \sigma^2_X)$ and observations $Y_1, Y_2, \ldots, Y_{n_2}$ come from $N(\mu_Y, \sigma^2_Y)$. Suppose that all observations are independent and that we are willing to assume that $\sigma^2_X = \sigma^2_Y$ which will then be denoted by σ^2. In this context, consider the problem of testing H_0: $\mu_X = \mu_Y$ against H_1: $\mu_X \neq \mu_Y$. This is indeed far from being an unrealistic problem. Descriptions of two situations of this type follow: (1) Students are assigned to one of two teachers at random. The teachers apply different instruction methods. At the conclusion of the course, the hypothesis of equality of population mean performances is tested in order to say whether or not the observed difference in the classes is statistically significant. (2) Newborn calves are assigned one of two rations at random. After three weeks, it is desired to test equality of means in order to say whether or not observed weight gains for the two rations are significantly different.

Letting $\delta = \mu_X - \mu_Y$, the hypothesis-testing problem H_0: $\delta = 0$ versus H_1: $\delta \neq 0$ is seen to be equivalent to the original problem. After deciding on n_1, n_2, and α, the suggested procedure is to reject H_0 if $(\bar{X} - \bar{Y})/s_p \sqrt{1/n_1 + 1/n_2} \geq$ tabulated t depending on α and $n_1 + n_2 - 2$ degrees of freedom, reject H_0 if $(\bar{X} - \bar{Y})/s_p \sqrt{1/n_1 + 1/n_2} \leq$ tabulated $-t$, and accept H_0 if $-t \leq (\bar{X} - \bar{Y})/s_p \sqrt{1/n_1 + 1/n_2} \leq t$.

A partial explanation of why this procedure was chosen follows. Since $\bar{X} \sim N(\mu_X, \sigma^2/n_1)$ and $\bar{Y} \sim N(\mu_Y, \sigma^2/n_2)$ then, by applying Theorems 3.1 and 3.2, we can say that

$$(\bar{X} - \bar{Y}) \sim N\left[\delta, \sigma^2\left(\frac{1}{n_1} + \frac{1}{n_2}\right)\right].$$

When H_0 is true, $\delta = 0$ and the standardized normal variable is

$$\frac{\bar{X} - \bar{Y}}{\sigma \sqrt{\dfrac{1}{n_1} + \dfrac{1}{n_2}}}.$$

One very important reason for choosing the pooled s^2_p to be our estimator of σ^2 is that, when we replace σ by s_p, the random variable $(\bar{X} - \bar{Y})/s_p \sqrt{1/n_1 + 1/n_2} \sim$ Student's t with $(n_1 + n_2 - 2)$ degrees of freedom.

The test described here is called a two-tailed Student's t test. A large difference in $(\bar{X} - \bar{Y})$ relative to $s_p \sqrt{1/n_1 + 1/n_2}$, whether it be negative or positive, demands, in agreement with intuition, the rejection of the statement $\mu_X - \mu_Y = 0$. A situation where H_1 is $\mu_X > \mu_Y$ calls for a one-tailed test where rejection of H_0 is the preferred action only if $(\bar{X} - \bar{Y})/s_p \sqrt{1/n_1 + 1/n_2}$ is greater than the tabulated t value. Similarly, when H_1 is $\mu_X < \mu_Y$, we reject H_0 only if

$$\frac{(\bar{X} - \bar{Y})}{s_p \sqrt{\dfrac{1}{n_1} + \dfrac{1}{n_2}}}$$

is less than the tabulated t value. Several problems found in the exercises to follow illustrate the various cases.

The statistic $(\bar{X} - \bar{Y})/s_p \sqrt{1/n_1 + 1/n_2}$ serves not only as a test statistic for testing hypotheses concerning means of two normal populations, but also as the statistic for deriving a 100γ percent confidence interval for the difference $\delta = \mu_X - \mu_Y$. Reading the appropriate tabulated values $\pm t(\nu, \gamma)$ from the t table, we can write

$$P\left[-t(\nu, \gamma) \le \frac{\bar{X} - \bar{Y} - \delta}{s_p \sqrt{\dfrac{1}{n_1} + \dfrac{1}{n_2}}} \le t(\nu, \gamma) \right] = \gamma.$$

In a manner similar to that used several times before, this probability statement can be worked into the probability statement

$$P\left[\bar{X} - \bar{Y} - t(\nu, \gamma)s_p \sqrt{\frac{1}{n_1} + \frac{1}{n_2}} \le \delta \le \bar{X} - \bar{Y} + t(\nu, \gamma) s_p \sqrt{\frac{1}{n_1} + \frac{1}{n_2}} \right] = \gamma.$$

This 100γ percent confidence interval has variable length

$$2t(\nu, \gamma)s_p \sqrt{\frac{1}{n_1} + \frac{1}{n_2}}.$$

The expected length for intervals created by applying the above formula can be reduced by any combination of the following three devices: (1) decrease γ; (2) increase n_1; and (3) increase n_2.

In concluding this section, we iterate that the statements made apply to situations where samples come from normal population with equal variance and all observations are independent. The methods may be inappropriate for cases different from those described here. Indeed, in a later section, we shall study the case where experiments are devised in such a way that observations are dependent.

3.5   EXERCISES

In the following problems let $X_i \sim \text{NID}(\mu_1, \sigma^2)$, $Y_i \sim \text{NID}(\mu_2, \sigma^2)$ and let all X_i be independent of all Y_i.

(1)   With $\alpha = .05$, test $\mu_1 = \mu_2$ against $\mu_1 \neq \mu_2$ using the academic observations: $X_i = \{3, 2, 1\}$, $Y_i = \{4, 2\}$.

(2)   Set a 95 percent confidence interval on $\delta = \mu_1 - \mu_2$ using the following academic observations: $X_i = \{7, 4, 7\}$, $Y_i = \{1, 2, 6\}$.

(3)   Let $\alpha = .01$, $n_1 = 11$, $n_2 = 16$. Test $\mu_1 = \mu_2$ against $\mu_1 \neq \mu_2$ if $\Sigma(X_i - \bar{X})^2 = 40$, $\bar{X} = 9.3$, $\Sigma(Y_i - \bar{X})^2 = 41$, and $\bar{Y} = 5.3$.

(4)   Set a 99 percent confidence interval on $\delta = \mu_1 - \mu_2$ if $n_1 = 7$, $n_2 = 13$, $s^2_1 = 9$, $s^2_2 = 10$, $\bar{X} = 3.1$, and $\bar{Y} = 1.7$.

(5)   With $\alpha = .05$, test $\mu_1 = \mu_2$ against $\mu_1 > \mu_2$ using the observations $X_i = \{2, 6, 5, 3\}$, $Y_i = \{1, 1, 4\}$.

(6)   With $\alpha = .01$, test $\mu_1 = \mu_2$ against $\mu_1 < \mu_2$ using the observations $X_i = \{3, 2, 1\}$, $Y_i = \{4, 2\}$.

(7)   A laboratory-type problem: Four determinations of the pH of a soil sample were made with one type of electrode and five determinations of the pH of the same sample were made with a second type of electrode. The readings were:

Type I: 5.76, 5.74, 5.84, and 5.80.
Type II: 5.82, 5.87, 5.96, 5.89, and 5.93.

Write out the details for setting up a hypothesis-testing problem for the hypothesis that the two types of electrodes were measuring the same thing. State all assumptions made and perform the test.

(8)   A laboratory-type problem: In a health experiment involving sleeping habits, the following data were collected.

With 7 or more hours of sleep				With 6 or less hours of sleep		
34.6	36.2	37.2	34.1	30.2	31.8	30.0
31.9	33.7	36.0	35.2	34.2	37.0	31.0
33.8	31.9	35.4	30.1	38.4	39.2	30.7
39.7	33.7	35.8				31.8

Are the populations from which these observations were taken significantly different? Formulate the problem to answer the above question and then answer the question asked by using statistical methods.

(9)   Suppose that one person claims that $\mu_X > \mu_Y$ and a second person claims

that $\mu_Y > \mu_X$. In order to settle the argument, nine observations are taken from each population. These observations follow:

Sample from population X					Sample from population Y				
14.2	14.7	14.9	14.5	14.1	14.2	14.6	14.5	14.3	14.1
14.8	14.6	14.7	14.6		14.2	14.6	14.4	14.0	

Formulate and then solve a statistical problem which might help to decide whether or not the observed difference is significant.

3.6   THE SNEDECOR *F* DISTRIBUTION AND SOME APPLICATIONS

In Exercise (8) of Section 3.5 the reader was asked if two populations were significantly different. There are, of course, many ways populations can differ, but if consideration is restricted to the normal family, populations are the same if variances and means coincide. Hence, in order to answer the question for normal populations, we need only to test equality of means and equality of variances. As yet we have not discussed a test for equality of variances. In preparation for a discussion of such tests, another family of distributions derived from the normal family will be discussed. The family will be referred to as the *Snedecor F family* in honor of G. Snedecor of Iowa State University who made great use of the distributions. The letter *F* is used in respect for Sir Ronald A. Fisher who was responsible for the derivation. The distribution, which is indexed by two parameters, has been tabulated in Table IV. It has application both in hypothesis problems involving equality of means and in hypothesis problems concerned with equality of variance.

Consider two independent variables U and V where $U \sim \chi^2(k)$ and $V \sim \chi^2(l)$. The nonnegative random variable

$$W = \frac{U}{k} \cdot \frac{l}{V}$$

is defined to be a Snedecor *F* variable with k and l degrees of freedom. We write $W \sim F(k, l)$ where it is understood that the k degrees of freedom for the Chi Square variable in the numerator is always mentioned before the l degrees of freedom for the denominator variable. An example of a Snedecor *F* variable that will presently be used in a hypothesis-testing situation is $W = \sigma^2_Y s^2_X / \sigma^2_X s^2_Y$. This statement is made with the restriction that s^2_X and s^2_Y are computed from normal random samples and that they are independent. If the sample size of the X population is n_1

and the sample size of the Y population is n_2, then the proof that $\sigma^2_Y s^2_X / \sigma^2_X s^2_Y \sim F(n_1 - 1, n_2 - 1)$ follows from the fact that

$$U = (n_1 - 1) \frac{s^2_X}{\sigma^2_X} \sim \chi^2(n_1 - 1)$$

and

$$V = (n_2 - 1) \frac{s^2_Y}{\sigma^2_Y} \sim \chi^2(n_2 - 1).$$

Forming the proper ratio, we have

$$\frac{U}{n_1 - 1} \cdot \frac{n_2 - 1}{V} = \frac{\sigma^2_Y s^2_X}{\sigma^2_X s^2_Y}.$$

The density curve for a Snedecor F variable has a general shape somewhat like a Chi Square curve. A change in either or both of the parameters k and l causes a change in the density curve. As mentioned earlier, only nonnegative values are taken on by a Snedecor F variable and the distribution is always skewed to the right.

Now focus on the problem of testing $\sigma^2_X = \sigma^2_Y$ against $\sigma^2_X \neq \sigma^2_Y$ where the test is to be based upon two independent normal random samples $X_i \sim N(\mu_X, \sigma^2_X)$, $i = 1, 2, \ldots, n_1$ and $Y_i \sim N(\mu_Y, \sigma^2_Y)$, $i = 1, 2, \ldots, n_2$. Intuitively, if the ratio s^2_X / s^2_Y is very large, one would be inclined to reject H_0, and if s^2_X / s^2_Y is very close to zero, one would likewise feel as if the null hypothesis should be rejected. This is in agreement with the conventional test procedure. After deciding on n_1, n_2, and α, accept H_0 if

Lower-tail tabulated   Upper-tail tabulated

$$F(n_1 - 1, n_2 - 1) \text{ value} < \frac{s^2_X}{s^2_Y} < F(n_1 - 1, n_2 - 1) \text{ value}$$

Depending on α Depending on α

and reject H_0 otherwise. Notice that when H_0 is true $\sigma^2_Y s^2_X / \sigma^2_X s^2_Y$ reduces to s^2_X / s^2_Y which then has a Snedecor F distribution. Procedures for one-sided alternative hypotheses follow in the same way they have for other situations.

Attention is now turned to another application of the Snedecor F tables. The concept presented here will be generalized in Chapter 4. Consider the Student t statistic

$$\frac{\bar{X} - \bar{Y}}{s_p \sqrt{1/n_1 + 1/n_2}}$$

which arose in connection with testing equality of means. If we square
this random variable, we have

$$W = \frac{(\bar{X} - \bar{Y})^2}{s^2_p(1/n_1 + 1/n_2)}.$$

It is relatively easy to show that $W \sim F(1, n_1 + n_2 - 2)$, as indeed it
can be shown that the square of any Student $t(\nu)$ variable is a Snedecor
$F(1, \nu)$ variable. Applying this fact to hypothesis testing, it follows that
two-sided symmetric t tests with ν degrees of freedom are equivalent to
one-sided upper-tail F tests with 1 and ν degrees of freedom.

Consider the following illustrative two-population normal situation.
Let $\sum_{i=1}^{10} (X_i - \bar{X})^2 = 90$, $\bar{X} = 17$, $\sum_{i=1}^{13} (Y_i - \bar{Y})^2 = 132$, and $\bar{Y} = 21$.
To test $\mu_X = \mu_Y$ against $\mu_X \neq \mu_Y$ with $\alpha = .05$, via a t test we assume
$\sigma^2_X = \sigma^2_Y$ (a reasonable assumption in this case), compute

$$\frac{\bar{X} - \bar{Y}}{s_p \sqrt{1/10 + 1/13}} = -2.9,$$

and compare it with the tabulated $t(21) = -2.08$. If the procedure is
to use an F test, we compute

$$\frac{(\bar{X} - \bar{Y})^2}{s^2_p(1/10 + 1/13)} = 8.39$$

and compare it with the tabulated $F(1, 21) = 4.33$. In either case, we
clearly reject the null hypothesis. In reading the tabulated values, be
aware that the positive t value is read with probability .975 to the left
while the F value is read with probability .95 to the left. Notice too that
the tabulated F value is the square of the tabulated t value.

3.7 EXERCISES

In the following exercises, consider $X_i \sim \text{NID}(\mu_X, \sigma^2_X)$, $i = 1, 2, \ldots, n_1$,
and $Y_i \sim \text{NID}(\mu_Y, \sigma^2_Y)$, $i = 1, 2, \ldots, n_2$.

(1) With $\alpha = .1$, test $\sigma^2_X = \sigma^2_Y$ against $\sigma^2_X \neq \sigma^2_Y$ when $X_i = \{1, 3, 2\}$ and
$Y_i = \{6, 10\}$. Can we test equality of means in this situation?

(2) With $\alpha = .05$, test $\sigma^2_X = \sigma^2_Y$ against $\sigma^2_X \neq \sigma^2_Y$ when $\Sigma X^2_i = 304$, $\Sigma X_i = 48$, $\Sigma Y^2_i = 1432$, $\Sigma Y_i = 140$, $n_1 = 8$, and $n_2 = 14$. Can we test equality
of means in this situation?

(3) With $n_1 = 11$, $n_2 = 16$, $\alpha = .01$, $\Sigma(X_i - \bar{X})^2 = 40$, and $\Sigma(Y_i - \bar{Y})^2 = 41$, test $\sigma^2_X = \sigma^2_Y$ against $\sigma^2_X > \sigma^2_Y$.

(4) With $\alpha = .05$, $\Sigma X^2_i = 134$, $\Sigma X_i = 42$, $\Sigma Y^2_i = 159$, $\Sigma Y_i = 48$, $n_1 = 21$, and $n_2 = 16$, test $\sigma^2_X = \sigma^2_Y$ against $\sigma^2_X < \sigma^2_Y$.

(5) Let $n_1 = 10$ and $n_2 = 12$. Use tables to obtain

(a) $P\left[\dfrac{(\bar{X} - \mu_{\bar{x}})\sqrt{n_1}}{s_X} \geq 2.262\right]$

(b) $P\left[\dfrac{(n_1 - 1)s^2_X}{\sigma^2_X} \geq 5.9\right]$

(c) $P\left[-2.262 < \dfrac{(\bar{X} - \mu_X)\sqrt{n_1}}{s_X} < 2.262\right]$

(d) $P\left[\dfrac{(n_1 - 1)s^2_X}{\sigma^2_X} < 16.9\right]$

(e) $P\left[\left|\dfrac{(\bar{X} - \mu_X)\sqrt{n_1}}{s_X}\right| \geq 1.383\right]$

(f) $P\left[2.70 \leq \dfrac{(n_1 - 1)s^2_X}{\sigma^2_X} \leq 19.0\right]$

(g) $P\left[\dfrac{\sigma^2_Y s^2_X}{\sigma^2_X s^2_Y} > 2.9\right]$

(h) $P\left[\dfrac{\sigma^2_X s^2_Y}{\sigma^2_Y s^2_X} > 3.1\right]$

(i) $P\left[.417 < \dfrac{\sigma^2_Y s^2_X}{\sigma^2_X s^2_Y} < 2.27\right]$

(j) $P\left[\dfrac{\sigma^2_Y s^2_X}{\sigma^2_X s^2_Y} < .986\right]$

(6) A laboratory-type problem: A long-standing process for making plastic has for years had a certain characteristic of its product described quite adequately by a member of the normal family of distributions. One day the process broke down. After repair, the product seemed erratic. Does the recorded data indicate a change in the distribution of the plastic characteristic? Support your answer with two F tests.

Before breakdown				After breakdown			
17.6	19.5	14.6	20.9	18.6	20.7	25.0	23.0
13.8	18.7	15.6	19.0	20.1	15.9	17.7	14.8
21.8	24.0	19.8	17.5	21.3	22.9	19.7	20.6
15.6	15.8	18.6					

(7) Eighteen plots of ground were used in an experiment involving two varieties of corn. The plots were chosen because they were found in the past to be homogeneous plots of ground. Variety X of corn was planted on nine plots chosen at random while variety Y was planted on the remaining nine plots. Assume that the following yields come from theoretical populations adequately described by normals and assume that the variances are equal. Test with an F test the hypothesis of equality of means.

X yields			*Y yields*		
87.6	81.9	80.4	81.3	76.6	78.6
75.9	78.6	73.8	71.9	74.6	70.0
82.6	80.5	85.4	83.5	77.8	76.5

(8) Derive a 95 percent confidence interval statement for the ratio σ^2_Y/σ^2_X and use the statement to set confidence intervals on σ^2_Y/σ^2_X, using the data in exercises (1) to (4).

3.8 PAIRED EXPERIMENTS AND THE CONCEPT OF EXPERIMENTAL DESIGN

Until the post-World War II period, advice to experimenters often took the following form: Hold all factors constant except the factor of interest. This advice is not given nearly so often today because experimenters now are interested generally in the combined effect of a change in several factors and because experimenters have convinced themselves that, in most studies, it is impossible to hold all except one factor constant. Admitting that the above remarks are true, the fact still remains that experiments often yield better data when experimental units are similar in many respects. This is basically why the psychologist studies twins, the agronomist selects homogeneous soil plots, the animal scientist chooses animals of about the same age and size, and the educator attempts to find students with similar backgrounds.

Perhaps the simplest experiments in which an attempt has been made to provide homogeneity of experimental units are *paired experiments*. In these experiments, two treatments or methods are studied by assigning at random one treatment to an experimental unit and the other treatment to another experimental unit very much like the first. Pairs of units need not be similar but units within a pair are hopefully very much alike. The data studied is usually the within-pair difference in response. To illustrate the idea, consider an experiment involving two methods of freezing meat. Suppose that the observations are tenderness measurements where low readings are the desirable numbers. To provide experimental units, homogeneous within pairs, chickens are separated left side from right side. This undoubtedly accomplishes the objective because we know of no reason why one side of a chicken should be consistently less tender than the other. Suppose that method X of freezing chicken is applied to one half of each chicken leaving method Y for each of the remaining halves. With proper identification applied, the meat is left stored for a prescribed period of time. Consider the following data after uniform thawing where the pair of numbers (X_i, Y_i) are the tenderness scores for the i chicken.

i	X_i	Y_i	$D_i = X_i - Y_i$
1	2.51	2.49	.02
2	1.62	1.55	.07
3	3.71	3.68	.03
4	2.80	2.81	−.01
5	2.02	1.98	.04
6	3.18	3.18	.00
7	1.94	1.90	.04

The advantage of pairing immediately becomes apparent. The tenderness of the thawed meat is variable to the extent that the effect of the method of freezing would be disguised in most experiments where the chickens or halves of chickens are assigned one of the two treatments in a completely random fashion. To illuminate the student on this point, consider the probable conclusion if, by chance, both halves of chickens 3 and 6 had been assigned method Y.

To study a concrete hypotheses-testing problem, suppose that the population mean δ of the conceptual population of all D_i is hypothesized to be $\delta = 0$. Suppose that we wish to test this hypothesis against $\delta \neq 0$. The null hypothesis is of course equivalent to the conjecture that the typical tenderness using freezing method X is the same as the typical tenderness using freezing method Y. If we assume that $D_i \sim N(\delta, \sigma^2_D)$, the best procedure is a t test. Experience has shown that differences are often adequately described by a member of the normal family even in situations where observation X and Y are poorly described by members of the normal family. There are also theoretical reasons to support the empirical findings. The t test consists of comparing the value of the test statistic $\bar{D}/s_{\bar{D}}$ with the tabulated values t and $-t$ read from the Student $t(n-1)$ table at the choosen probability values. Here $s_{\bar{D}} = s_D/\sqrt{n}$ and n is the number of pairs. It should be pointed out that, had we not paired results, the proper test statistic for the hypothesis $\delta = 0$ would have been $(\bar{X} - \bar{Y})/s_p \sqrt{2/n}$ with $2(n - 1)$ degrees of freedom. Notice that, since $\bar{D} = \bar{X} - \bar{Y}$, this test statistic could have been written $\bar{D} \sqrt{n}/s_p \sqrt{2}$. If we compute s_p and compare its value with that of s_D, we shall find that, in the illustrative example, the loss of $(n - 1)$ degrees of freedom by pairing has more than been compensated for, by the dramatic reduction in the standard deviation with which $\bar{D}$ is compared. This attempt to design an experiment in such a way that the difference of sample means is compared with a small standard deviation illustrates what is, perhaps, the basic principle underlying experimental design. Put in another way, we strive to compare differences of sample means with the standard deviation of experimental units that have been treated alike. The variance of experimental units that have been treated alike is often referred to as the experimental error variance.

3.9 EXERCISES

(1) Work out the details for testing $\delta = 0$ when $\alpha = .05$ in the illustrative paired-frozen-chicken example. Compute the value of $\sqrt{2}\, s_p$ in order to compare this with s_D. Compare also the tabulated critical values for $t(n - 1)$ and $t(n - 2)$.

(2) Use the following paired data to run a two-sided t test for equality of means at the $\alpha = .01$ level.

X Value: 10.9 8.6 7.0 9.8 5.9.
Y Value: 9.7 8.3 6.5 9.7 5.3.

Use the same data to test the same hypothesis with a one-sided F test at the $\alpha = .01$ level.

(3) Use the following paired data to test $\mu_X = \mu_Y$ against $\mu_X > \mu_Y$ with α chosen to be .05.

X Value: 97 82 90 75.
Y Value: 91 81 91 72.

(4) A laboratory-type problem: A physical education teacher in a certain large high school wanted to study frequency of exercise. He took advantage of the unusual event of having ten sets of twin boys in his high school. The teacher decided to have one twin from each set lift weights twice a week and the other twin in each set lift weights four times a week. After several weeks the improvement for each boy, in pounds lifted, was recorded as below.

Twins	2/week	4/week	Twins	2/week	4/week
A	21	23	F	14	18
B	9	14	G	32	39
C	37	35	H	15	14
D	29	30	I	22	27
E	15	21	J	20	29

Test the hypothesis of equality of means for the two frequencies of exercise using $\alpha = .05$.

(5) Work out the derivation details for setting a 100γ percent confidence interval on the parameter $\delta = \mu_X - \mu_Y$ when data comes from a paired experiment.

(6) Compute a 95 percent confidence interval estimate of $\delta = \mu_X - \mu_Y$ using the following data pairs:

X: 27 19 31 21 16.
Y: 26 22 29 20 14.

(7) How can the expected length of a 100γ percent confidence interval for the difference δ be reduced when data is paired?

(8) Determine the approximate sample size needed in order that the probability that $\bar{D}$ differs from δ by less than three units be about .95 if the D_i are normal and $\sigma^2_D = 36$.

4

STATISTICAL INFERENCE RELATIVE TO SAMPLING FROM MORE THAN TWO NORMAL POPULATIONS

4.1 INTRODUCTION TO THE STUDY OF MORE THAN TWO POPULATIONS

In earlier chapters, we have studied experimental situations where two treatments, two varieties, or two methods were under consideration, thereby creating a two-population situation to be analyzed. In this chapter attention ultimately will be directed toward applying techniques for testing hypotheses, setting confidence intervals, and determining point estimators in situations where observations come from more than two normal populations. Most of the new concepts will be presented by example. As a preview of things to come, consider the situation where three varieties of wheat are grown with the intention of testing the null hypothesis that there exists equality among all mean yields for the varieties; that is, if μ_i is the mean yield for the conceptual population of the ith variety, then the null hypothesis is $H_0: \mu_1 = \mu_2 = \mu_3$. For rather obvious reasons, a t test is not suitable for this situation. Consideration of this fact provides motivation for new and, in some ways, more general methods for testing hypotheses. Before presenting a test procedure for handling the above situation, some related aspects of the situation will be investigated.

When many populations are being considered, the use of k different letters to represent observations from k different populations does not work out very effectively. Much more convenient is the notation Y_{ij} used to represent the j observation in the ith population. Thus Y_{32} would denote the second observation in the third population and not the 32nd observation. To avoid confusion, with integers greater than nine we would write, for example, $Y_{11,24}$ to represent the 24th observation in the 11th population. Careful attention is called to other notation. $Y_{i.}$ will represent the total of observations from the ith population, and $Y_{..}$ will represent the total of all observations from all populations.

Notation is somewhat different when populations are classified in more than one way. We need to discern carefully between two cases illustrated by the following situations. A soldier is classified as belonging to the first squad of the second platoon. This is *nested* or *hierarchal classification*. Days of a year are classified according to both humidity and temperature. This is *cross-classification*. Oranges on branches on trees in groves on farms in counties illustrates hierarchal classification. Boys with blue eyes, blond hair, and height over six feet illustrates cross-classification. Later, when two-way classification is discussed, considerations will first be directed toward cross-classification. At that time the notation Y_{ijk} will be associated with the kth observation in category i of one classification and category j of the second classification. If one and only one observation is taken from the (i, j) category pair, the subscript k will be deleted and Y_{ij} will denote the observation.

In this chapter are found introductions to the subject of formal analysis of variance and the concept of a statistical model. A general observation or two seems appropriate at this time. *Analysis of variance*, which will henceforth be abbreviated as AOV, in one sense is the filling out of a more or less standard format after an experiment has been run. The format, properly filled out, serves as a useful source of numerical information. AOV techniques, which are relatively standard for a wide variety of situations, avoid the necessity of memorizing many calculation formulas. As for a statistical model, it can be thought of as an attempt to explain observations gathered from the real world.

In the discussion of analysis of variance, the expression "reduction in sum of squares due to considering the mean" will often be used. An explanation of this expression will be given at this time. Consider n observations $Y_1, \ldots, Y_n$. It has earlier been shown that

$$\sum_{i=1}^{n} (Y_i - \bar{Y})^2 = \sum_{i=1}^{n} Y_i^2 - n\bar{Y}^2.$$

Other ways of writing this with y_i denoting $Y_i - \bar{Y}$ are

$$\sum_{i=1}^{n} y_i^2 = \sum_{i=1}^{n} Y_i^2 - n\bar{Y}^2 \quad \text{and} \quad n\bar{Y}^2 = \sum_{i=1}^{n} Y_i^2 - \sum_{i=1}^{n} y_i^2.$$

The last expression exhibits $n\bar{Y}^2$ as the reduction in the sum of squares $\sum_{i=1}^{n} Y^2_i$ due to subtracting $\bar{Y}$ from each of the observations Y_i. In order to further exemplify this point, consider the observations $Y_1 = 5$, $Y_2 = 1$, and $Y_3 = 6$. Since $\bar{Y}$ for this data is 4, we have $y_1 = 1$, $y_2 = -3$, and $y_3 = 2$. The sums of squares $\sum_{i=1}^{n} Y^2_i$ and $\sum_{i=1}^{n} y^2_i$ are 62 and 14, hence the reduction in sum of squares is the difference, namely 48, and this is precisely the value of $n\bar{Y}^2 = 3(4)^2$.

4.2 THE COMPLETELY RANDOMIZED EXPERIMENTAL DESIGN

Consider an experimental situation where three varieties of wheat were each grown on four equal-area plots of ground. Suppose that the plots were assigned to the varieties at random, resulting in a field layout as shown in Figure 4.1, where varieties are denoted by Roman numerals

I	2.6	III	3.3	I	2.7	II	2.1		
III	3.2	III	2.3	I	2.8	II	2.2	II	2.0
				II	2.5	III	3.0	I	2.5

Figure 4.1

and yields are recorded in hundreds of bushels. In this experiment, observations are classified one way and the design is called *completely randomized* (CRD). The same data arranged in what is called a statistical layout appears in Table 4.1. Suppose that the experiment was run and data compiled in order to test equality of means.

Table 4.1 A Statistical Layout for a CRD

Variety	I	II	III	
Yields	$Y_{11} = 2.6$	$Y_{21} = 2.5$	$Y_{31} = 3.2$	
	$Y_{12} = 2.7$	$Y_{22} = 2.1$	$Y_{32} = 3.3$	
	$Y_{13} = 2.8$	$Y_{23} = 2.2$	$Y_{33} = 2.3$	
	$Y_{14} = 2.5$	$Y_{24} = 2.0$	$Y_{34} = 3.0$	
Totals	$Y_{1.} = 10.6$	$Y_{2.} = 8.8$	$Y_{3.} = 11.8$	$Y_{..} = 31.2$
Averages	$\bar{Y}_1 = 2.65$	$\bar{Y}_2 = 2.20$	$\bar{Y}_3 = 2.95$	

The question concerning equality of means can be asked in other ways. Is there a significant difference among the totals Y_i? Has evidence been gathered which is strong enough to reject the hypothesis at the $\alpha = .05$ level? Is the variability among the totals Y_i. significantly greater than the variability within varieties? The concept of comparing variability among varieties with variability within varieties is our departure point for an excursion in the world of formal analysis of variance. Hopefully, the trip will be exciting with many interesting side trips. Part of the AOV format for the data of our illustration is displayed in Table 4.2.

Table 4.2   An AOV for Varieties

Source of variation	Degrees of freedom	Sum of squares	Mean square
Total	$rt = 12$	$\sum_i \sum_j Y^2_{ij} = 83.06$	
Reduction due to the mean	1	$n\bar{Y}^2 = 81.12$	
Between varieties	$(t-1) = 2$	$\dfrac{1}{r}\sum_{i=1}^{t} Y^2_{i.} - n\bar{Y}^2 = 1.14$	.57000
Pooled within varieties	$t(r-1) = 9$	$\sum_{i=1}^{t}\sum_{j=1}^{r}\left(Y_{ij} - \dfrac{Y_{i.}}{r}\right)^2 = .80$	.08888

Not much justification is given at the present time for the way in which the various columns of the format are filled in. The letter t is used to denote the number of variates (the word treatment is usually used when considering the general situation) and r is used to denote the number of observations per variety.

The sum of squares calculations will now be more carefully explained.

$$\text{Total } SS = \sum_i \sum_j Y^2_{ij} = (2.6)^2 + (2.7)^2 + \cdots + (2.3)^2 + (3.0)^2 = 83.06$$

Reduction due to the mean (also called correction factor) $= n\bar{Y}^2$
$$= 12(2.6)^2 = 81.12$$

$$\text{Between varieties } SS = \frac{(10.6)^2}{4} + \frac{(8.8)^2}{4} + \frac{(11.8)^2}{4} - n\bar{Y}^2 = 1.14$$

$$\text{Within variety } SS = (2.6 - 2.65)^2 + (2.7 - 2.65)^2 + \cdots + (3.0 - 2.95)^2 = .80.$$

Notice that the degrees of freedom 9, 2, and 1 add to equal the sample size of 12. This is not an accident because, in terms of t and r, we have $1 + (t - 1) + t(r - 1) = rt$. Furthermore, the sum of squares column is such that the total SS is the sum of the sum of squares associated with the other sources of variation. These are features of all AOV's. Indeed, an AOV has been defined to be a procedure for partitioning the total sum of squares into parts associated with the sources of variation. One component sum of squares can be obtained by subtracting the rest from the total SS. In completely randomized experiments, the pooled-within-variety SS is usually obtained by subtraction.

The mean-square column of numbers is obtained by dividing each sum of squares by the associated degrees of freedom. Notice that, in so doing, the pooled-within-variety mean square plays a role similar to that played by s^2_p and the mean square for between varieties is equivalent to a sample variance for varieties. If Exercises (3) and (4), which follow, are worked correctly it will be found that the ratio of mean squares for between and within varieties takes on the squared value of the t statistic for testing equality of means when two varieties are studied, normal and independent observations result, and equality of variances is assumed. Although there is no t test counterpart, the ratio of between to within mean squares is the preferred test statistic when more than two populations are considered, because when all populations have the same population variance and all population means are equal, the ratio of between to within mean squares is distributed as Snedecor $F[t - 1, t(r - 1)]$. The test of equality of means then consists of comparing the ratio obtained from the AOV with the proper tabulated critical Snedecor F value.

4.3 EXERCISES

(1) Display a statistical layout for the following field-layout data. Suppose the data to be normal data associated with three varieties of corn. Write out an AOV associated with this academic data. Assume equality of variances and test equality of means with $\alpha = .05$.

						I 4
I 7	III 7	II 9	III 6	I 4	III 4	I 3
II 8	II 9	I 4	III 4	III 5	II 8	II 6

(2) Write out an AOV for the following statistical layout. Compute the sum of squares for within treatments in two ways.

Treatment	I	II	III	IV
	1.1	1.2	1.1	1.4
	1.0	1.3	1.1	1.2

(3) Based on the following normal data and assuming equality of variances for the populations, test equality of means using the t test procedure presented in Chapter 3.

Treatment	I	II
	7	6
	6	3
	7	3
	4	4

(4) Write out the AOV for the data in Exercise (3) and compute the ratio of the mean squares for between and within varieties. Compare this ratio with the test statistic computed in Exercise (3).

(5) Determine the ratio of the mean squares for between and within treatments for the following data.

Treatment	I	II	III
	7	8	7
	6	9	6
	4	9	4
	4	8	5
	3	6	4

(6) A laboratory-type problem: Four feeds were fed to lots of five baby chicks. The gains in weights were

Feed	I	II	III	IV
	5.5	6.1	4.2	16.4
	4.8	10.2	9.7	12.2
	4.2	4.5	8.1	16.1
	2.7	8.9	9.5	6.5
	5.3	6.3	9.2	9.3

Compare the mean squares for among lots and within lots.

4.4 THE CONCEPT OF A LINEAR STATISTICAL MODEL

Consider the common variance, three-population situation where $Y_{1j} \sim N(\mu_1, \sigma^2)$, $Y_{2j} \sim N(\mu_2, \sigma^2)$, $Y_{3j} \sim N(\mu_3, \sigma^2)$, and all Y_{ij} are independent. What follows is an attempt to explain the observation Y_{ij}. Let $e_{ij} = Y_{ij} - \mu_i$, then $Y_{ij} = \mu_i + e_{ij}$ and $e_{ij} \sim \text{NID}(0, \sigma^2)$. If, in addition to this, we let $\mu = \frac{1}{3} \sum_{i=1}^{3} \mu_i$ and $v_i = \mu_i - \mu$, we can then write $Y_{ij} = \mu + v_i + e_{ij}$. This is an example of a *linear statistical model* with the following properties:

(1) μ is a fixed unknown parameter.

(2) The v_i are fixed unknown parameters (in our illustration, they are the variety effects).

(3) $\sum_{i=1}^{3} v_i = 0$.

(4) $e_{ij} \sim \text{NID}(0, \sigma^2)$.

It should be pointed out that there is one component in the model for each row in the AOV. The model is clearly associated with the completely randomized experiment and it is called a *fixed-effects model* because the v_i's were assumed to be fixed parameters.

To help anchor ideas related to the concept of a linear model, values of the parameters μ, v_1, v_2, v_3, and σ^2 will be assumed and some academic observations created from them by applying the fixed effects completely randomized one-way classification model. Suppose $\mu = 5$, $v_1 = 3$, $v_2 = -1$, $v_3 = -2$, and $\sigma^2 = 4$. Reading random normal numbers from Table VI, we obtain for the random component the rounded-off numbers

$$e_{11} = -0.9 \qquad e_{21} = 1.9 \qquad e_{31} = 1.5$$
$$e_{12} = -2.9 \qquad e_{22} = -0.9 \qquad e_{32} = 0.7.$$

The sample mean and sample variance for these errors are $\bar{e} = -.1$ and $s^2_e = 3.264$. The observations $Y_{ij} = \mu + v_i + e_{ij}$ along with the totals and averages for the varieties are

$Y_{11} = 5 + 3 - .9 = 7.1$	$Y_{21} = 5 - 1 + 1.9 = 5.9$	$Y_{31} = 5 - 2 + 1.5 = 4.5$
$Y_{12} = 5 + 3 - 2.9 = 5.1$	$Y_{22} = 5 - 1 - .9 = 3.1$	$Y_{32} = 5 - 2 + .7 = 3.7$

Totals	$Y_{1.} = 12.2$	$Y_{2.} = 9.0$	$Y_{3.} = 8.2$
Averages	$\bar{Y}_1 = 6.1$	$\bar{Y}_2 = 4.5$	$\bar{Y}_3 = 4.1$

The created data will be used next to obtain unbiased estimates of the parameters with which we started. In real-world problems we seldom, if ever, have the opportunity to compare the actual values of parameters with estimates of the parameters. Keep in mind that the estimates in this illustration are based on very few degrees of freedom and notice too that s^2_e is less than the actual value of $\sigma^2 = 4$; hence, it cannot be said that our sample of errors is wild in the sense of having excessive variability. It also might be noted that the value of the average e_{ij} supports the assumption that the errors have a population mean of zero.

$$\hat{\mu} = \frac{Y_{..}}{6} = \frac{29.4}{6} = 4.9$$

whereas $\mu = 5$.

$$\hat{\mu}_1 = \bar{Y}_1 = 6.1, \qquad \hat{\mu}_2 = \bar{Y}_2 = 4.5, \qquad \hat{\mu}_3 = \bar{Y}_3 = 4.1,$$

$$\hat{v}_1 = \hat{\mu}_1 - \hat{\mu} = 1.2, \qquad \hat{v}_2 = \hat{\mu}_3 - \hat{\mu} = -.4, \qquad \hat{v}_3 = \hat{\mu}_3 - \hat{\mu} = -.8,$$

while

$$v_1 = 3.0, \qquad v_2 = -1.0, \qquad v_3 = -2.0.$$

The estimate s^2_e for σ^2 based on five degrees of freedom cannot be computed from the observations. There is, however, another estimate of σ^2 based on three degrees of freedom which is easily obtained from the AOV (Table 4.3).

Table 4.3   An AOV for the Illustrative Created Data

Source of variation	df	SS	MS
Total	6	154.78	
Reduction due to the mean	1	144.06	
Between varieties	2	4.48	2.24
Within varieties	3	6.24	2.08

The mean square for within varieties (that is, 2.08) is an unbiased estimate of σ^2 based on three degrees of freedom. Like s^2_e it underestimates, in this illustration, the actual variance of the population from which we sampled.

In concluding the illustration, attention is called to the fact that, although the population means were selected as different, the data created do not supply evidence to strongly suggest that the means are different. The computed F value is near one whereas the tabulated $F(2, 3)$ with $\alpha = .05$ is 9.55. Our conclusion then is that the error variance has disguised the differences in the populations.

The fixed-effects model discussed here is not the only linear, additive model used to describe observations coming from a one-way classification situation. Consider the linear, additive relationship $Y_{ij} = \mu + \tau_i + \epsilon_{ij}$ with the properties that

(1) μ is a fixed parameter,
(2) $\tau_i \sim \text{NID}(0, \sigma^2_\tau)$,
(3) $\epsilon_{ij} \sim \text{NID}(0, \sigma^2_\epsilon)$,
(4) τ_i and ϵ_{ij} are independent for all i and j.

This is called the *random-effects*, one-way classification, linear, additive model. It is especially applicable to situations where inference to a large number of treatments, varieties, or methods is to be based on a sample of the treatments, varieties, or methods. The model is associated with a very special case of hierarchal classification of experimental units but the concepts considered here form the basis for the analysis of all hierarchal designs. When the variances σ^2_τ and σ^2_ϵ are the objects of attention, this model is sometimes called a *variance-components model*.

To illustrate the computation of estimates of variance components, consider the following illustration familiar to all animal-science experimenters (Table 4.4). Let Y_{ij} be the weight gain in a fixed-time interval

Table 4.4

Litter	I	II	III	
	5.0	4.9	6.0	
Weight	4.5	6.1	6.7	
Gains	4.2	6.0	5.9	
	3.9	5.8	6.6	
Totals	17.6	22.8	25.2	65.6

for the jth pig in the ith litter. Suppose that the variance-components model $Y_{ij} = \mu + l_i + p_{ij}$ is adopted where $l_i \sim \text{NID}(0, \sigma^2_l)$ and $p_{ij} \sim \text{NID}(0, \sigma^2_p)$. Let the problem be the estimation of σ^2_l and σ^2_p. The AOV

Table 4.5

Source	df	SS	MS	$\mathcal{E}MS$
Total	12	368.2200		
Mean	1	358.6133		
Between litters	2	7.5467	3.7733	$\sigma^2_p + 4\sigma^2_l$
Within litters	9	2.0600	.2289	σ^2_p

provides the calculations necessary for an easy determination of these parameters. Fortunately this easy method provides estimates with good properties, some of which will be considered later. The procedure is to set computed mean squares equal to expected means (Table 4.5) and solve the resulting equations for $\hat{\sigma}^2_l$ and $\hat{\sigma}^2_p$. For the data in the illustration, we set

$$\hat{\sigma}^2_p + 4\hat{\sigma}^2_l = 3.7733$$

$$\hat{\sigma}^2_p = .2289$$

from which the estimates $\hat{\sigma}^2_p = .2289$ and $\hat{\sigma}^2_l = .8861$ are obtained. The proper coefficient for σ^2_l in the case where each population has r observations is the value r.

4.5 ONE-WAY CLASSIFICATION EXPERIMENTS WITH UNEQUAL NUMBERS PER CLASSIFICATION

Consider the following statistical layout for one-way classification data where Y_{ij} represents the jth observation in classification i.

Classification	I	II	III
	Y_{11}	Y_{21}	Y_{31}
	Y_{12}	Y_{22}	Y_{32}
	.	.	.
	.	.	.
	Y_{1n_1}	.	Y_{3n_3}
		.	
		.	
		Y_{2n_2}	
Totals	$Y_{1.}$	$Y_{2.}$	$Y_{3.}$

With t representing the number of classifications and $n. = \sum_{i=1}^{t} n_i$ denoting the total number of experimental units, the appropriate AOV details are given in Table 4.6.

This design, which is frequently used because of its flexibility, is unusual in that most calculations are not much more difficult with unequal numbers per classification than they are for the case where each classification possesses the same number of experimental units. An aspect in which there exists an appreciable difference in the calculations

Table 4.6   A Random-Effects AOV

Source	df	SS	$\mathcal{E}MS$ for random model
Total	$n.$	$\displaystyle\sum_{i=1}^{t}\sum_{j=1}^{n_i} Y^2{}_{ij}$	
Mean	1	$Y^2{}_{..}/n.$	
Between classifications	$t-1$	$\displaystyle\sum_{i=1}^{t} (Y^2{}_{i.}/n_i) - Y^2{}_{..}/n.$	$\sigma^2{}_\epsilon + k\sigma^2{}_\tau$
Within classifications	$\displaystyle\sum_{i=1}^{t} (n_i - 1)$	$\displaystyle\sum_{i=1}^{t}\sum_{j=1}^{n_i} Y^2{}_{ij} - \sum_{i=1}^{t} (Y^2{}_{i.}/n_i)$	$\sigma^2{}_\epsilon$

is the mean square for between classifications. If a random-effect model, $Y_{ij} = \mu + t_i + \epsilon_{ij}$, is assumed, the between classification expected mean square is $\sigma^2{}_\epsilon + k\sigma^2{}_t$ where

$$k = \frac{n^2. - \displaystyle\sum_{i=1}^{t} n^2{}_i}{n.(t-1)}.$$

With either a fixed-effects model or a random-effects model describing the data, the appropriate test for equality of classification means is to compare the ratio of between to within classification mean squares with the proper tabulated $F\left[t-1, \displaystyle\sum_{i=1}^{t} (n_i - 1) \right]$ value.

4.6   EXERCISES

(1) Use the AOV method to estimate the variance components $\sigma^2{}_\epsilon$ and $\sigma^2{}_\tau$ when the model $Y_{ij} = \mu + \tau_i + \epsilon_{ij}$, $\tau_i \sim \text{NID}(0,\ \sigma^2{}_\tau)$, and $\epsilon_{ij} \sim \text{NID}(0,\ \sigma^2{}_\epsilon)$ is applied to the data:

Temperature	I	II	III	IV	V
Observations	1.7	1.8	1.3	1.8	1.4
	1.6	1.9	1.5	1.8	1.6

(2) Test the hypothesis of equality of population means by first writing out an AOV for the following data. State the assumptions needed in order that the test be exact with probability of a Type I error equal to .05.

Population	I	II	III
	2.0	2.8	2.8
	2.6	2.2	2.8
		2.6	2.4
		2.4	

(3) Using the random-effects model $Y_{ij} = \mu + \tau_i + \epsilon_{ij}$ with $\tau_i \sim \text{NID}(0, 4)$, $\epsilon_{ij} \sim \text{NID}(0, 1)$, and $\mu = 10$, create five observations for each of four populations by reading values of τ_i and ϵ_{ij} at random from Tables V and VI. After creating the 20 observations, write out a one-way classification AOV for the created data and estimate, μ, σ^2_τ and σ^2_ϵ.

(4) Associating the fixed-effects model $Y_{ij} = \mu + \tau_i + \epsilon_{ij}$ with the following data, estimate unbiasedly the parameters $\mu_1, \tau_1, \ldots, \tau_t$, and σ^2 if σ^2 is the variance of $\epsilon_{ij} \sim \text{NID}(0, \sigma^2)$ and $\sum_{i=1}^{t} \tau_i = 0$.

Treatment	I	II	III
	13	18	19
	15	15	21
	18	21	22

(5) Apply the Student t test methods of Chapter 3 to the following data to test equality of treatment means.

X: 1.6, 7.9, 1.2, 6.3, 5.8, 4.3, 9.5, 3.6, 8.7, 5.5, 6.8, 6.1.
Y: 3.2, 6.2, 5.0, 1.3, 2.8, 4.4, 1.2, 3.3.

(6) For the data in Exercise (5) write out the model, including all parameter and random variable assumptions, and write out the AOV. From the AOV compute the F ratio and compare it with the Student t test statistic of Exercise (5).

(7) A laboratory-type problem: The data in this problem come from a completely randomized experiment conducted in order to study the effect of certain treatments on the yield of early cabbage.

Treatment	I	II	III	IV	V	VI
	14.07	13.64	12.86	9.14	10.58	9.14
	13.93	13.68	13.04	11.15	11.79	9.73
	11.85	15.10	11.15	11.34	11.64	9.44
	13.45	13.02	11.52	10.37	9.20	10.32

Assume the model to be $Y_{ij} = \mu + \tau_i + \epsilon_{ij}$ where μ is a fixed parameter representing in a sense the typical yield of all cabbage of the variety grown, $\epsilon_{ij} \sim \text{NID}(0, \sigma^2)$ and τ_i is a constant equal to the effect of treatment "i" as measured from the mean μ.

(a) Estimate $\alpha_1, \alpha_2, \ldots, \alpha_6$ where $\alpha_i = \mu + \tau_i$.
(b) Write out the AOV for this experiment.
(c) Estimate σ^2 from the AOV calculations.
(d) With probability of a Type I error equal to .05, test the hypothesis $\alpha_1 = \alpha_2 + \cdots + \alpha_6$ against the alternative hypothesis that there exists at least one inequality among the α_i's.

(8) A laboratory-type problem: Consider the following high-school math-test scores for geometry students from seven high schools in an area where the school boards work together and all students use the same materials, including tests.

School	A	B	C	D	E	F	G
	89.6	85.5	70.5	82.9	86.0	82.5	86.3
	84.0	91.6	78.8	80.5	75.3	92.4	87.4
	80.5	84.3	89.5	70.8	68.4	80.4	74.3
	90.1	73.2	82.3	77.5	88.3	81.0	69.4
	72.8	78.5		82.5			89.0
	83.8			88.4			

Test the hypothesis of equal school means, stating the model and all assumptions pertaining thereto.

(9) A laboratory-type problem: Three greenhouses were used in an experiment conducted with chrysanthemum plants as experimental units to help determine the relative magnitudes of the greenhouse and plant components of variance. From the following data, estimate σ^2_h and σ^2_p, assuming the model $Y_{ij} = \mu + h_i + p_{ij}$ with $h_i \sim \text{NID}(0, \sigma^2_h)$ and $p_{ij} \sim \text{NID}(0, \sigma^2_p)$.

House	I	II	III
	21.3	20.4	19.9
	19.8	20.0	19.1
	20.7	20.9	20.3
Observations	21.0	21.9	20.4
	20.9	22.3	20.1
	19.7	19.9	19.2
	20.5	20.5	20.5

(10)   Prove the following algebraic identities used in writing out sums of squares for AOV's.

(a) $$\sum_{i=1}^{2} \left(X_i - \frac{X_.}{2} \right)^2 = \sum_{i=1}^{2} X^2_i - \frac{X^2_.}{2} = \frac{(X_1 - X_2)^2}{2}$$

(b) $$\sum_{i=1}^{2} \frac{Y^2_{i.}}{r} - \frac{Y^2_{..}}{2r} = \frac{1}{2r}(Y_{1.} - Y_{2.})^2$$

(c) $$\sum_{i=1}^{t} \sum_{j=1}^{r} Y^2_{ij} = \left[\sum_{i=1}^{t} \frac{Y^2_{i.}}{r} - \frac{Y^2_{..}}{rt} \right]$$

$$+ \left[\sum_{i=1}^{t} \sum_{j=1}^{r} Y^2_{ij} - \sum_{i=1}^{t} \frac{Y^2_{i.}}{r} \right] + \frac{Y^2_{..}}{rt}$$

4.7 TWO-WAY CROSS-CLASSIFICATION DESIGNS

In this section attention is directed toward a slightly more complicated situation. There are many reasons for studying what is discussed here, not the least of which is the fact that it helps the student to understand better the one-way classification situation. The concepts are approached in story fashion. Consider a circumstance where four researchers each claim to have developed a superior variety of tomato for growth in Pennsylvania. The director of the research agrees to provide a dozen test plots for growing the tomatoes. Suppose that the test plots are in an area between a highway and a river where it is known from past experience that the plots near the river are more fertile than those along the highway with a noticeable gradient in fertility as one moves away from the highway. A completely random assignment of plots might result in an unfair statistical layout such as shown in Figure 4.2.

The researcher with variety IV would justifiably complain about such an assignment of plots. Observations from such a design would have variety effects confounded with fertility effects. To avoid unfair assignments of plots, the area could be partitioned into three subareas which will be called *blocks*. Within each block the varieties are then assigned to the homogeneous plots at random, resulting in what is referred to as a *randomized block design*. The design is a special case of the more general class of two-way cross-classification designs. A field layout and a corresponding statistical layout along with some academic observations for a three-block, four-variety randomized block design are displayed in Figure 4.3 and Table 4.7.

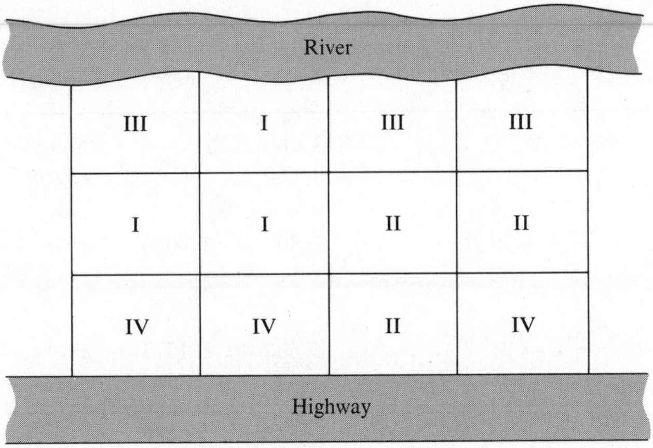

Figure 4.2

Field layout

II	9.9	IV	9.0	I	9.7	III	8.6
III	7.8	IV	8.3	II	9.1	I	7.9
I	8.0	III	6.6	IV	6.5	II	8.2

Figure 4.3

Table 4.7 Statistical Layout

| | | | Variety | | | |
		I	II	III	IV	Totals
	I	9.7	9.9	8.6	9.0	37.2
Block	II	7.9	9.1	7.8	8.3	33.1
	III	8.0	8.2	6.6	6.5	29.3
Totals		25.6	27.2	23.0	23.8	99.6

The AOV for these data and an explanation of the computation of the block and variety sums of squares follows (see Table 4.8). The residual sum of squares is obtained by subtraction while the total and mean SS are obtained in the conventional manner.

Letting Y_{ij} represent the observation for the jth variety in block i, the

$$\text{Block } SS = \sum_i \frac{Y^2_{i.}}{t} - \frac{Y^2_{..}}{bt} = \frac{(37.2)^2 + (33.1)^2 + (29.3)^2}{4} - 826.68$$

$$= 7.805$$

Table 4.8   AOV

SV	df	SS	MS	Expected Mean
Total	$bt = 12$	839.060		Squares for fixed-
Mean	1	826.680		effects model
Blocks	$b - 1 = 2$	7.805	3.9025	$\sigma^2 + \dfrac{t}{b-1}\sum_{i=1}^{b} \beta^2{}_i$
Varieties	$t - 1 = 3$	3.533	1.1777	$\sigma^2 + \dfrac{b}{t-1}\sum_{i=1}^{t} \tau^2{}_j$
Residual	$(b-1)(t-1) = 6$	1.042	.1737	σ^2

and

$$\text{Variety } SS = \sum_{j} \frac{Y^2{}_{.j}}{b} - \frac{Y^2{}_{..}}{bt} = \frac{(25.6)^2 + (27.2)^2 + (23.0)^2 + (23.8)^2}{3}$$

$$- 826.68 = 1.1777.$$

Before considering some of the uses for the entries in an AOV for a randomized block experiment, consideration will be given to a model frequently used in conjunction with two-way cross-classification data. Let the expected value of the observation Y_{ij} for variety j in block i be denoted by μ_{ij} and assume that for each pair (i, j) the observation $Y_{ij} \sim$ NID(μ_{ij}, σ^2). Letting $\epsilon_{ij} = Y_{ij} - \mu_{ij}$ and assuming the linear additive relationship $\mu_{ij} = \mu + \beta_i + \tau_j$ with $\sum_{i=1}^{b} \beta_i = 0$ and $\sum_{j=1}^{t} \tau_j = 0$, we can write $Y_{ij} = \mu + \beta_i + \tau_j + \epsilon_{ij}$. This is an example of a linear additive model with the properties:

(1) μ is a fixed unknown parameter.
(2) β_i and τ_j are fixed unknown parameters (in our illustration they are the block and variety effects).
(3) $\sum_{i=1}^{b} \beta_i = 0, \sum_{j=1}^{t} \tau_j = 0$.
(4) $\epsilon_{ij} \sim$ NID$(0, \sigma^2)$.

Again there is one component in the model for each row in the AOV. Alternative models that are in common use are the random-effects model, where β_i and τ_j are normal and independent random variables, and the mixed model, where the β_i's or the τ_j's are independent random variables and the others are fixed parameters.

Linear additive models have been found to adequately describe many randomized block design situations. For each of the three models described above, the mean square for residual has expected value equal to σ^2. Other expected values for the fixed-effects model are shown in the AOV. These mean squares can be used to argue that the F test involving the ratio of mean square of varieties to mean square for residual is a good test procedure for testing the null hypothesis of equal variety effects. Indeed, it can be shown that when each τ_j is zero, this ratio has a Snedecor $F[t - 1, (t - 1)(b - 1)]$ distribution. When the variety effects τ_j are very different, the mean square for varieties has an expected value equal to $\sigma^2 + \dfrac{b}{t - 1} \displaystyle\sum_{j=1}^{t} \tau^2_j$. This will be greater than the expected value of the mean square for residual and thus the ratio of mean squares will tend to take on values greater than one. On the other hand, when the variety effects τ_j are each zero, MSV and MSR estimate the same value σ^2 and their ratio tends to hover near one. The test of the hypothesis of equal variety effects consists of comparing the proper, upper-tailed, tabulated, $F[(t - 1), (t - 1)(b - 1)]$ value with the ratio MSV/MSR. The hypothesis is rejected when the computed value exceeds the tabulated value.

Consider again the illustration. The computed F ratio for testing the hypothesis of equal variety effects is about 6.78 whereas the tabulated $F(3, 6)$ value when $\alpha = .05$ is 4.76. As evidenced by the size of the mean square for blocks, the premise that blocks are different in fertility seems to be well founded. Had the block source of variability not been considered, the degrees of freedom and sum of squares for blocks would have been pooled with those for residual thereby greatly inflating the mean square for residual and, in effect, disguising variety effects, if there were any.

4.8 EXERCISES RELATED TO THE TWO-WAY CROSS-CLASSIFICATION TYPE OF EXPERIMENT

(1) Consider the following field layout:

I	A 4	C 5	B 4
Block II	B 2	A 3	C 3
III	A 2	C 3	B 1

(a) Display a statistical layout for this data.

(b) Write a fixed-effects linear additive model for this data.

(c) Write out an AOV for the data.

(d) Do you think the difference in treatments A, B, and C would have been declared significant if a completely randomized experiment had been run?

(2) Write out an AOV for the following two-way classification data, and write a fixed-effects linear additive model for the data.

Treatment	I	II	III	IV
Block I	2.1	2.0	2.3	2.5
Block II	1.3	1.5	1.7	1.7

(3) Apply the Student t test methods of Chapter 3 to the following paired data to test equality of treatment means.

Pair	I	II	III	IV	V	VI	VII
Treatment I	8.7	9.9	7.1	5.9	7.8	8.3	8.9
Treatment II	8.1	8.9	6.9	6.1	7.6	8.0	8.2

(4) For the data in Exercise (3) write out the linear additive paired-observation fixed-effects model. Write out the AOV for the two-way cross-classification data using each pair as a block. From the AOV compute the F ratio and compare it with the Student t test statistic of Exercise (3).

(5) A laboratory-type problem: Twenty-four students were grouped according to previous grade averages in mathematics. Three sections of a new course were set up and taught by the same teacher but with three different books, methods, and materials. The observation recorded was the average grade for each student in the new course. Analyze these two-way classification data stating all assumptions needed in your analysis.

Previous grade average:	95–100	90–95	85–90	80–85	75–80	70–75	65–70	60–65
Method, book, and materials I	90.3	92.8	90.0	85.7	75.9	70.8	78.5	71.0
Method, book, and materials II	87.9	86.5	84.6	70.1	70.0	65.3	66.3	65.4
Method, book, and materials III	94.5	84.7	84.5	75.8	80.5	66.2	60.9	70.4

(6) A laboratory-type problem: A strength characteristic for items coming from an industrial process was measured for 20 items that had been classified according to curing time and machine on which they were made.

The data were as follows:

Curing time		I	II	III	IV	V
Machine	I	40.8	42.7	49.8	43.5	39.7
	II	39.7	40.3	46.3	43.4	39.9
	III	40.5	41.7	42.9	43.8	40.6
	IV	41.6	44.8	45.8	46.0	44.3

Assume the model $Y_{ij} = \mu + m_i + \tau_j + \epsilon_{ij}$ where μ is a fixed parameter (representing in a sense the typical characteristic for the process), m_i is the fixed effect of the i machine, τ_j is the fixed effect of the j time, and $\epsilon_{ij} \sim \text{NID}(0, \sigma^2)$. Let $\sum_{i=1}^{4} m_i = 0$ and $\sum_{j=1}^{5} \tau_j = 0$ since it is always possible to define the parameters in such a way that this holds.

(a)   Estimate each $\alpha_i = \mu + m_i$.

(b)   Estimate each $\gamma_j = \mu + \tau_j$.

(c)   Write out the AOV for this experiment.

(d)   Estimate σ^2 from the AOV calculations.

(e)   Test with probability of a Type I error equal to .05 the hypothesis $\alpha_1 = \alpha_2 + \alpha_3 = \alpha_4$ against the alternative hypothesis that there exists at least one inequality among the α_i's.

(f)   Test with probability of a Type I error equal to .05 the hypothesis $\gamma_1 = \gamma_2 = \gamma_3 = \gamma_4 = \gamma_5$ against the alternative hypothesis that there exists at least one inequality among the γ_j's.

(7)   Using the mixed model $Y_{ij} = \mu + \beta_i + \tau_j + \epsilon_{ij}$ with $\mu = 8$, $\tau_1 = 2$, $\tau_2 = -1$, $\tau_3 = -1$, $\beta_i \sim \text{NID}(0, 4)$, and $\epsilon_{ij} \sim \text{NID}(0, 1)$ create 12 observations by reading random values of β_i and ϵ_{ij} from Tables VI and V. After creating the observations, write out an AOV for the data and estimate from the data the parameters μ, τ_1, τ_2, τ_3, σ_β^2, and σ_ϵ^2.

5

BASIC CONCEPTS WHEN TWO CHARACTERISTICS ARE STUDIED

5.1 INTRODUCTION

Consider experimental units for which the values of two characteristics are measured, yielding pairs of observations (X, Y). Associated with pairs of observations are the concepts of covariance, linear correlation, linear regression, and least squares. There are many applications that could be cited but mention is made of two at this time.

(a) Let (X_i, Y_i) be the IQ test score and the college index for person i.

(b) Let (U_i, V_i) be the rainfall in May and the corn yield in June for the ith year in Oklahoma.

Roughly speaking, regression theory relates to problems where one characteristic is of principal interest and the other is measured because knowledge of its value may shed light on the distribution of values for the first characteristic. Correlation theory, on the other hand, usually pertains to problems where the relationship between X and Y is the object of study. Correlation is a much used, sometimes misused concept. In a sense, this chapter could be said to deal with infinitely many populations, since we could consider a different population of Y's for each value of X or vice versa. In another sense, this chapter deals with one bivariate distribution.

5.2   THE CONCEPT OF COVARIANCE AND THE CONCEPT OF LINEAR CORRELATION

The statistic

$$S_{XY} = \sum_{i=1}^{n} \frac{(X_i - \bar{X})(Y_i - \bar{Y})}{n - 1}$$

was considered briefly in Chapter 3 in connection with two normal populations. We now want to consider S_{XY} in the context of pairs of observations (X_i, Y_i) for experimental units $i = 1, 2, \ldots, n$. S_{XY} is called the *sample covariance* of X and Y while $\sum_{i=1}^{n} (X_i - \bar{X})(Y_i - \bar{Y})$ is referred to as the *sum of corrected cross products*, the word corrected being used to denote the fact that we are considering deviations. Now, whereas S_{XY} depends on the scale of measurement, the statistic

$$r = \frac{S_{XY}}{S_X S_Y}$$

is free of the scale and is an often-used measure of the degree of association for X and Y. Although r will be referred to as the correlation, it is more accurately described as the *sample linear correlation coefficient* of the variables X and Y.

In certain contexts r can be thought of as a sample estimate of a corresponding bivariate population parameter

$$\rho = \mathcal{E}\left[\left(\frac{X - \mu_X}{\sigma_X}\right)\left(\frac{Y - \mu_Y}{\sigma_Y}\right)\right]$$

In general, r is not an excellent estimator of ρ. Two reasons for this are (1) r is variable except for large values of the sample size and (2) except when $\rho = 0$, r is a biased estimate of ρ. Despite these drawbacks, r is often used as a point estimator of ρ. Interval estimation procedures for ρ and tests of hypotheses related to ρ exist but will not be discussed at this time. We end our brief discussion of ρ with the important remark that, when random variables X and Y are independent, ρ is zero, but when ρ is zero the variables X and Y are not necessarily independent.

As a notational device, throughout this chapter we shall let $x_i = X_i - \bar{X}$ and, similarly, for other variables we shall denote by little letters the deviations from the sample mean of the characteristic.

Several of the elementary properties of the sample correlation coefficient r will be found in the following set of exercises.

5.3 EXERCISES

In the following exercises, whenever pairs of numbers are presented, the first number in each pair is to be considered the value of the X characteristic and the second number the value of the Y characteristic.

(1)   Plot the data $(1, 3)$, $(2, 6)$, $(1, 4)$, $(4, 7)$, and compute r.

(2)   Plot the data $(1, 5)$, $(4, 4)$, $(6, 6)$, $(2, 7)$, $(2, 8)$, and compute r.

(3)   Plot the data $(5, 2)$, $(3, 6)$, $(1, 8)$, $(3, 4)$, and compute r.

(4)   Compute r when the data is $(1, 2)$, $(3, 6)$, $(2, 4)$. Plot the data and note that $Y_i = 2X_i$.

(5)   Prove that $r = 1$ whenever $Y_i = cX_i$, $i = 1, 2, \ldots, n$, and $c > 0$.

(6)   Prove that $r = -1$ whenever $Y_i = dX_i$, $i = 1, 2, \ldots, n$, and $d < 0$.

(7)   Using any one of the above sets of data, compute the value of S_{XY}. Now for the set of data chosen, let $Z_i = 3X_i$ and compute the values of S_{ZY} and the correlation coefficient of Z and Y.

(8)   For the data in Exercise (1), compute $\displaystyle\sum_{i=1}^{4} X_i y_i$, $\displaystyle\sum_{i=1}^{4} x_i Y_i$, and $\displaystyle\sum_{i=1}^{4} x_i y_i$.

(9)   Prove in general that for any set of data $\Sigma x_i y_i = \Sigma X_i y_i$.

(10)   Do you think that the variables X and Y are related when observation pairs for these variables are $(0, 5)$; $(3, 4)$; $(-4, 3)$; $(5, 0)$; $(3, -4)$; $(0, -5)$; $(-5, 0)$; $(-3, 4)$; $(-4, -3)$; $(-3, -4)$; $(4, 3)$; $(4, -3)$? Plot the data and compute the value of r.

(11)   Show that $-1 \leq r \leq 1$ whenever $(\Sigma x_i y_i)^2 \leq (\Sigma x^2_i)(\Sigma y^2_i)$. *Comment:* It can be shown that the condition always holds; hence the sample correlation coefficient always has the property that $-1 \leq r \leq 1$.

5.4 INTRODUCTION TO LEAST SQUARES

Consider pairs of observations (X_i, Y_i) $i = 1, 2, \ldots, n$ where the variable Y is the variable of principal interest. The variable X will be referred to as the covariable or auxiliary variable. In most texts it is called the independent variable. In considering observation pairs, men of science and business years ago observed the fact that, in many situations, the points (X_i, Y_i) tend to fall on or near some straight line. It seemed reasonable and later it became important to fit a straight line to data. The ruling in of a good-looking line was for many situations a sufficiently good procedure but for some situations it was not. Many criterions can and have been proposed for a "best" straight line for a set of data but eventually the criterion of *least squares* showed itself to have more good properties than other criterions for almost all situations.

Rather than pursue the topic of least squares in a general setting, the ideas will be presented through the following academic example. Suppose that numerical values were recorded for the same two character-istics for each of five experimental units. Designating by Y the value of the characteristic of principal interest, let the data be

X: 1   2   3   4   5.
Y: 1.2   1.0   1.9   3.4   3.0.

Figure 5.1 shows the location of the five pairs of observation and sug-gests, perhaps, the existence of a linear trend.

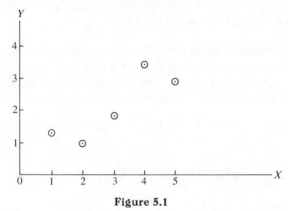

Figure 5.1

There are many, in fact an uncountable number of lines, that could be fit to the data. Among criterions that might be considered are the following:

(1)   Select a line that minimizes the maximum vertical deviation.
(2)   Select a line that minimizes the sum of the vertical deviations.
(3)   Select a line that minimizes the maximum perpendicular deviation.
(4)   Select a line that minimizes the sum of the perpendicular deviations.

The data of Figure 5.1 are reproduced in Figure 5.2 along with a line that illustrates the third criterion. The student is asked to draw lines that adequately illustrate the satisfying of the other criterions.

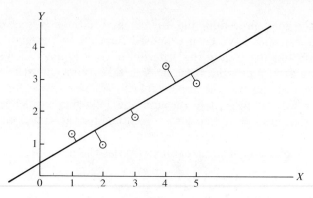

Figure 5.2

Although these criterions and others have proved useful in special problems, it is the following criterion that has proved to be, by far, the most useful criterion in most situations.

The Least Squares Criterion Consider a set of n points (X_i, Y_i) in two-dimensional space. The line $f(X) = a + bX$ which best fits the data in the least squares sense is the line that minimizes the sum of squares of vertical deviations. To express the concept in symbols, let $d_i = Y_i - f(X_i)$ and let $Z = \sum_{i=1}^{n} d^2{}_i$; then a least squares fit is the line that minimizes

$$Z = \sum_{i=1}^{n} [Y_i - f(X_i)]^2 = \sum_{i=1}^{n} (Y_i - a - bX_i)^2.$$

Let us first minimize Z with respect to a, holding b at some arbitrary but fixed value. Z can be written as a quadratic in the variable a.

$$Z = \Sigma(Y_i - bX_i - a)^2 = \Sigma[a^2 - 2a(Y_i - bX_i) + (Y_i - bX_i)^2]$$

$$Z = n\left[a^2 - \frac{2}{n} a\Sigma(Y_i - bX_i)\right] + \Sigma(Y_i - bX_i)^2.$$

Completing the square with respect to a gives

$$Z = n\left[a - \frac{1}{n}\Sigma(Y_i - bX_i)\right]^2 + \Sigma(Y_i - bX_i)^2 - \frac{[\Sigma(Y_i - bX_i)]^2}{n}$$

and substituting $\bar{Y}$ and $\bar{X}$ for $(1/n)\Sigma Y_i$ and $(1/n)\Sigma X_i$, respectively, we have

$$Z = n[a - (\bar{Y} - b\bar{X})]^2 + \Sigma(Y_i - bX_i)^2 - \frac{[\Sigma(Y_i - bX_i)]^2}{n}.$$

Z can be minimized with respect to a for any fixed b by choosing $a = \bar{Y} - b\bar{X}$.

Employing this value of a, we write

$$Z = \Sigma(Y_i - bX_i - \bar{Y} + b\bar{X})^2 = \Sigma(y_i - bx_i)^2.$$

Completing the square with respect to b, we have

$$Z = \Sigma x^2_i \left[b^2 - \frac{2b\Sigma x_i y_i}{\Sigma x^2_i} + \frac{(\Sigma x_i y_i)^2}{(\Sigma x^2_i)^2} \right] + \Sigma y^2_i - \frac{(\Sigma x_i y_i)^2}{\Sigma x^2_i}$$

$$Z = \Sigma x^2_i \left(b - \frac{\Sigma x_i y_i}{\Sigma x^2_i} \right)^2 + \Sigma y^2_i - \frac{(\Sigma x_i y_i)^2}{\Sigma x^2_i}.$$

This form shows that Z is minimum when $b = (\Sigma x_i y_i/\Sigma x^2_i)$ and that the minimum is $\Sigma y^2_i - (\Sigma x_i y_i)^2/(\Sigma x^2_i)$.

Calculations of a and b for the example are found in Table 5.1. From

Table 5.1   Least Squares Computations

X_i	Y_i	x_i	x^2_i	$x_i Y_i$	$\bar{Y} + bx_i =$ $a + bX_i$	$d_i =$ $y_i - bx_i$	d^2_i
1	1.2	-2	4	-2.4	.9	.3	.09
2	1.0	-1	1	-1.0	1.5	$-.5$	.25
3	1.9	0	0	0	2.1	$-.2$	.04
4	3.4	1	1	3.4	2.7	.7	.49
5	3.0	2	4	6.0	3.3	$-.3$	.09
Totals 15	10.5	0	10	6.0	10.5	0	.96

the calculations exhibited in the table we obtain, for our illustration, the values $b = \Sigma xY/\Sigma x^2 = \frac{6}{10} = .6$ and $a = \bar{Y} - b\bar{X} = .3$. The straight line $f(X) = .3 + .6X$ is plotted in Figure 5.3 along with the observation points and the vertical deviations from the observation points to the line.

A line created by the method described is often referred to as a regression line for the variable Y on the variable X. This terminology is of historical interest. In the late nineteenth century, F. Galton studied data of the type that we are considering here. Publishing in the Proceedings of the Royal Society of London, in papers such as his 1897 paper entitled "The Average Contribution of Each Several Ancestor to the Total Heritage of the Offspring," he used the term regression. He considered situations such as the following. Let (X_i, Y_i) be the ith observation pair where X_i is a father's height and Y_i is the adult height of his oldest male offspring. He found that although tall fathers tend to have

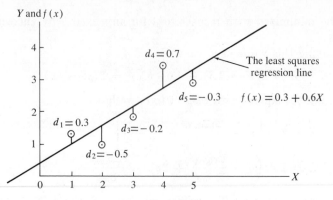

Figure 5.3

tall offspring, those offspring belonging to fathers of about the same excessive height have an average height less than that of their fathers. Galton referred to this phenomenon as regression back toward the mean of the population. For short fathers of about the same height, the offspring had an average height greater than that of the fathers and, furthermore, the scatter diagram of all pairs (X_i, Y_i) appeared to have a strong linear trend. The term *regression* has stayed, although today it has a far more general connotation, referring to a broad class of ideas and procedures in multivariate statistics.

Displayed in Table 5.1, in addition to the data and certain quantities useful in the calculation of the intercept a and the slope b, are columns of differences and squared differences. The differences $Y_i - (a + bX_i) = y_i - bx_i = d_i$ are geometrically displayed in Figure 5.3. The sum of squares $\sum_{i=1}^{5} d^2_i$ is for the illustration data found to be .96. It is emphasized here that this sum of squares, namely .96, of vertical deviations from observation points to the straight line is less than the corresponding sum of squares for any other straight line.

5.5 EXERCISES

(1)   Consider the data pairs $\begin{pmatrix} X \\ Y \end{pmatrix} = \begin{pmatrix} 0 \\ 5 \end{pmatrix}, \begin{pmatrix} 1 \\ 7 \end{pmatrix}, \begin{pmatrix} 2 \\ 6 \end{pmatrix}, \begin{pmatrix} 3 \\ 10 \end{pmatrix}$.

     (a)   Plot the observation points.
     (b)   Determine the least squares regression line $f(X) = a + bX$.
     (c)   Plot the least squares regression line.
     (d)   Calculate and picture on the graph the four deviations d_i, $i = 1$, . . . , 4.

(e) Compute $\sum\limits_{i=1}^{4} d^2_i$ and $\sum\limits_{i=1}^{4} y^2_i$.

(f) Check to see if the least squares line passes through the point $(\bar{X}, \bar{Y})$.

(2) Do the calculations asked for in Exercise (1) but use the data pairs (1, 2), (1, 0), (3, 4), and (3, 3).

(3) Prove that the least squares regression line always goes through the point $(\bar{X}, \bar{Y})$.

(4) d_i is defined to be $Y_i - f(X_i)$. Show that d_i is also equal to $y_i - bx_i$.

(5) Determine the least squares regression line for the data (X, Y): (3, 6); (1, 8); (4, 1). From the regression line, obtain a predicted value of Y when $X = 2$.

(6) A laboratory-type problem: Consider the following IQ and college index scores for 20 recent graduates of a college.

IQ	Index	IQ	Index	IQ	Index	IQ	Index
105	2.1	132	3.2	104	2.0	99	1.9
107	2.4	130	3.8	112	2.2	120	2.3
113	3.0	115	2.0	119	2.8	125	3.0
127	3.1	117	3.1	106	1.9	126	2.7
97	2.3	124	2.7	104	2.5	135	3.0

(a) Determine the least squares regression line for the regression of college index on IQ.

(b) Using the regression line, predict the college index for an incoming freshman with IQ equal to (1), 110; (2), 120; and (3), 130.

5.6 MATHEMATICAL MODELS FOR REGRESSION

Earlier in this text the concept of a population for a random variable Y was presented. In this section this concept is extended by considering populations of Y values for each of many different X values. Primarily the case to be studied will be where the values of X are fixed and chosen in advance of running the experiment, but later some comments will be made concerning the case where X and Y are both random variables.

For each chosen observable value X of an auxiliary characteristic, measured without error, we conceive of a population of Y values for a primary characteristic under study. These Y values, for fixed X, have a population mean that will be denoted by $\mu_{Y|X}$. Plotted in Figure 5.4 are values of X versus $\mu_{Y|X}$.

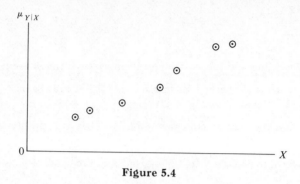

Figure 5.4

If, for each X in an interval $a \le X \le b$, a mean $\mu_{Y|X}$ is associated with X, then the locus of points $(X_i, \mu_{Y|X_i})$ might appear as in Figure 5.5.

The set of all points $(X_i, \mu_{Y|X_i})$ is referred to as the population curve for the regression of Y on X. Attention in this chapter is focused on the

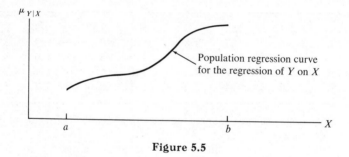

Figure 5.5

situation where the population regression curve is a straight line and the notation for this straight line will be

$$\mu_{Y|X} = \alpha + \beta X.$$

A straight-line regression curve along with the distribution of Y's for three values of X is pictured in Figure 5.6.

Unless the slope β equals zero, the distribution of Y for fixed X of course will not have the same mean as the mean for the distribution of Y associated with another value of X. In other respects the distributions may also differ. If indeed they belong to the same family, they may still have different variances. We study here a very special case, yet a case that has proved to be extremely useful. The distribution of Y for a given value of X is assumed to be normal, the variance of Y for a given value of X does not depend on the value of X, and all values of Y for a given value of X are assumed to be independent. These restrictions

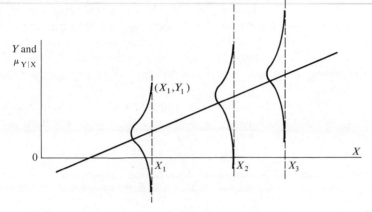

Figure 5.6

can be expressed concisely in symbols by writing for an observation pair (X_i, Y_i),

$$Y_i \sim \mathrm{NID}(\alpha + \beta X_i, \sigma^2).$$

If ϵ_i is used to express the deviation of Y_i from its mean $\mu_{Y|X}$, then $\epsilon_i = Y_i - \mu_{Y|X_i} = Y_i - \alpha - \beta X_i$ and, transposing, we can express the observation Y_i as

$$(*) \qquad Y_i = \alpha + \beta X_i + \epsilon_i.$$

It should be noted that σ^2 as used here is usually denoted by other authors with the more cumbersome but more enlightening notation $\sigma^2_{Y|X}$. Observe too that $\epsilon_i \sim \mathrm{NID}(0, \sigma^2)$. Equation (*) along with the properties of α, β, X_i and ϵ_i is referred to as a *linear regression model*. It is important to remember that a model is basically an attempt to explain data and that there are other models that can and are used in regression-type studies. The model described here is used because it adequately describes many situations and because it has mathematical properties that make it nice to work with. These properties remain nice as long as X is observable without error. In other words, it is important to be able to say that $Y|X \sim N(\alpha + \beta X, \sigma^2)$ and not have $Y|X \sim N(\alpha + \beta(X + \epsilon), \sigma^2)$.

The observation Y_i is to be thought of as composed of three additive parts. To α, the fixed part, is added a component that depends on the magnitude of the covariable X. To $\alpha + \beta X_i$ is added a component that explains why not all of the observation points are to be found on a straight line. The variable ϵ_i is referred to as the random-error contribution to the observation Y_i. The model perhaps may be better understood by employing as an aid an adequate geometrical mental image. Many and perhaps most of the early statisticians made frequent use of geometry to help think about and to convey statistical concepts. Consider a truck full of hot chocolate traveling along a straight road on a cold day. Picture the mound of chocolate created when suddenly the tail gate flies open, and for an interval of time, hot chocolate uniformly flows out onto the cold pavement where it solidifies. Its shape can be compared to that of an old-fashioned eaves trough turned upside down along the regression line. See Figure 5.6.

The model here described has three unknown parameters, α, β, and σ^2. Unbiased estimators of these parameters, which also possess many other nice properties, are: $\hat{\alpha} = a$, $\hat{\beta} = b$, and

$$\hat{\sigma}^2 = \sum_{i=1}^{n} \frac{d^2{}_i}{n-2} = \sum_{i=1}^{n} \frac{(Y_i - a - bX_i)^2}{n-2}.$$

For the illustrative example, the value of $\hat{\sigma}^2$ is $.96/3 = .32$.

In a sense the line $f(X) = a + bX$ is an estimate of the line $\mu_{Y|X} = \alpha + \beta X$. If the data does come from a population with a linear trend $\mu_{Y|X} = \alpha + \beta X$ and if least squares is the criterion used, then $f(X) = a + bX$ is the best fit of the data.

It may seem natural to work with the equation $f(X) = a + bX$ to represent a straight line and it was for that reason that the concept of regression was introduced with an equation of that form. However, it is more convenient to work with an alternative form of the straight-line equation. Since the least squares criterion yielded $a = \bar{Y} - b\bar{X}$ as the intercept on the Y axis, we can write for the least squares regression line $f(X) = \bar{Y} - b\bar{X} + bX$ or $f(X) = \bar{Y} + bx$. Likewise, in the equation for the true regression line, if we introduce the substitution $X = x + \bar{X}$ we can write $\mu_{Y|X} = \alpha + \beta x + \beta \bar{X}$ and upon replacing $\alpha + \beta \bar{X}$ by μ we have $\mu_{Y|X} = \mu + \beta x$. The observation Y_i can then be expressed in the following two ways:

(1) $Y_i = \bar{Y} + bx_i + d_i$, which follows from the fact that $d_i = Y_i - f(X_i)$;
(2) $Y_i = \mu + \beta x_i + \epsilon_i$.

An unbiased estimate of μ can be shown to be $\hat{\mu} = \bar{Y}$. It is for this reason that μ is often referred to as the overall population mean.

In dealing with the theoretical aspects of regression, considerations usually will be confined to the situation that has been presented thus far. However, there is another situation that can be described by the model $Y_i = \mu + \beta x_i + \epsilon_i$, $\epsilon_i \sim \text{NID}(0, \sigma^2)$. When the pair (X_i, Y_i) has a bivariate distribution, regression methods apply equally as well as when the X_i are controlled by the experimenter. The designation "model I" is often given to the case where the X_i are fixed and "model II" designates the case where the X_i and Y_i are random. In the latter case there are two regression lines (generally different) since we could consider the regression of X on Y as well as the regression of Y on X. The last statement is subject to the ever-present restriction that the auxiliary variable be observable without error. (Note that X can be observable without error and still be random.)

5.7   THE REGRESSION AOV AND TESTS OF HYPOTHESES

It has proved to be instructive, when working in the framework of least squares regression, to study variance from several points of view. The topic first will be investigated by examining algebraically the partitioning of the total sums of squares. As was seen earlier, an observation Y_i can be expressed as $Y_i = \bar{Y} + bx_i + d_i$ hence

$$Y^2_i = \bar{Y}^2 + b^2x^2_i + d^2_i + 2\bar{Y}bx_i + 2\bar{Y}\,d_i + 2bx_id_i$$

and

$$\sum_{i=1}^{n} Y^2_i = n\bar{Y}^2 + b^2 \sum_{i=1}^{n} x^2_i + \sum_{i=1}^{n} d^2_i + 2\bar{Y}b \sum_{i=1}^{n} x_i + 2\bar{Y} \sum_{i=1}^{n} d_i + 2b \sum_{i=1}^{n} x_id_i.$$

Since it can easily be shown that each of the last three terms sum to zero, we have

$$(**) \qquad \sum_{i=1}^{n} Y^2_i = n\bar{Y}^2 + b^2 \sum_{i=1}^{n} x^2_i + \sum_{i=1}^{n} d^2_i.$$

The statistic $n\bar{Y}^2$, in agreement with earlier notation, is called the reduction in sum of squares due to considering sum of squares of deviations about the mean instead of about zero. The statistic $\sum_{i=1}^{n} d^2_i$, as was noted earlier, is the sum of squares of deviations about the least squares regression line. The statistic

$$b^2 \sum_{i=1}^{n} x^2_i = \sum_{i=1}^{n} Y^2_i - n\bar{Y}^2 - \sum_{i=1}^{n} d^2_i = \sum_{i=1}^{n} (Y_i - \bar{Y})^2 - \sum_{i=1}^{n} d^2_i$$

must then be the reduction in sum of squares due to considering deviations about the regression line instead of deviations about the mean $\bar{Y}$. The situation is somewhat analogous to that of a working man taking home a pay check of $\sum_{i=1}^{n} Y^2_i$ dollars and finding, after his wife has spent $n\bar{Y}^2$ dollars and his daughter $b^2 \sum\limits_{i=1}^{n} x^2{}_i$ dollars, that his residual is

$\sum\limits_{i=1}^{n} d^2{}_i$ dollars. The analogy holds also in the respect that $\sum\limits_{i=1}^{n} d^2{}_i$ is usually

small relative to $\sum\limits_{i=1}^{n} Y^2{}_i$.

Referring now to the illustrative example considered in the first section, the geometry of the situation will now be studied. In Figure 5.7

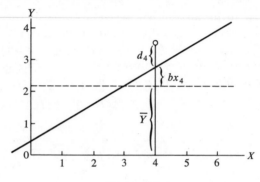

Figure 5.7

the magnitude of the observation Y_4 is pictured as the sum of the three
components $\bar{Y}$, bx_4, and d_4. Each Y_i can be pictured in a similar way if
we properly interpret the role of a negative quantity. Notice again that,
when the components $\bar{Y}$, bx_i, and d_i are squared and summed, they become
the three-component sum of squares for the total sum of squares. When
cross products vanish, such as in our situation, the sources of variability
associated with the partitioned sum of squares are said to be *orthogonal*.
Orthogonality is a desired property in statistical analyses and statistical
designs are often chosen on the basis of whether or not the analysis has
this property.

Analysis of variance when two characteristics are studied can conveniently be put into a format similar to that used for studies involving
one characteristic. In regression-type situations, it is customary to use a
slightly different notation. $R(\mu)$ will designate the reduction in sum of
squares due to considering deviations about $\bar{Y}$ instead of about zero and
$R(\beta|\mu)$ will denote the reduction in sum of square due to considering
deviations about the least squares regression line instead of deviations
about the mean $\bar{Y}$. The deviations from the linear regression line will be

referred to as residuals. The format for simple linear regression of Y on X follows:

Source of variation	Degrees of freedom	Sum of squares	Mean square
Total	n	$\sum\limits_{i=1}^{n} Y^2{}_i$	
$R(\mu)$	1	$n\bar{Y}^2$	$n\bar{Y}^2$
$R(\beta\|\mu)$	1	$b^2 \sum\limits_{i=1}^{n} x^2{}_i$	$b^2 \sum\limits_{i=1}^{n} x^2{}_i$
Residual	$n-2$	$\sum\limits_{i=1}^{n} d^2{}_i$	$\dfrac{\sum\limits_{i=1}^{n} d^2{}_i}{n-2}$

In reflecting upon the format it is observed that the sum of squares column is a vertically expressed statement of the result (**) and, as in other experimental design situations, the degrees of freedom column plays the role of a "bookkeeping" column. If we assume a linear regression model with $\epsilon_i \sim N(0, \sigma^2)$ then in one sense, the entire format is a convenient manner in which to set down the calculations for obtaining the unbiased estimate $\sum\limits_{i=1}^{n} d^2{}_i/(n-2)$ for σ^2. Presently the regression AOV will reveal itself as being helpful in other ways.

Consider now a sample of n from a population that is described by the linear regression model $Y_i = \mu + \beta x_i + \epsilon_i$, $\epsilon_i \sim NID(0, \sigma^2)$ for X_i which are fixed in the interval (a, b). From the n observation pairs, the value of the random variable

$$b = \hat{\beta} = \sum_{i=1}^{n} x_i Y_i \Big/ \sum_{i=1}^{n} x^2{}_i$$

can be computed. The distribution of the random variable b is our immediate interest.

We first observe that b is a linear combination of normal variables and is hence a normal variable. Since the expected value of Y_i is $\mu + \beta x_i$, the expected value of b works out to be β and the variance of b, denoted by $\sigma^2{}_b$, is equal to $\sigma^2 \Big/ \sum\limits_{i=1}^{n} x^2{}_i$. The details of deriving these results are left as exercises. The second of these results has significance with respect to designing an experiment for generating data from which a precise estimate of β can be computed. The population variance σ^2_b can be made small by making $\sum\limits_{i=1}^{n} x^2_i$ large. In addition to making the sample size large, the values of X_i should be chosen at or near the endpoints of the interval $[a, b]$ in order to enlarge the deviations x_i. The concept here is the same as that involved in constructing, with proper slope, a straight line by locating as accurately as possible two points through which the line is to pass. Obviously, two points separated by a distance almost equal to the length of the ruling instrument give a better result than two points chosen close to each other. The concept can perhaps be made more meaningful by associating it with the likely consequences of driving an auto with a very tiny steering wheel.

Summarizing the facts, we can write

$$b \sim N\left(\beta, \frac{\sigma^2}{\sum\limits_{i=1}^{n} x^2_i}\right).$$

The situation with regard to the distribution of

$$\hat{\sigma}^2 = \frac{\sum\limits_{i=1}^{n} d^2_i}{n-2}$$

is equally interesting. It is well known and easily shown using the theory of quadratic forms that $\sum\limits_{i=1}^{n} (d_i/\sigma)^2 \sim \chi^2(n-2)$ and is independent of b. Hence $\hat{\sigma}^2/\sigma^2$ is the ratio of a $\chi^2(n-2)$ variable to its degrees of freedom. This fact will be used often in setting up tests of hypotheses and working out interval estimates for regression parameters.

With the distribution facts in hand, we are now in a position to test the null hypothesis $H_0: \beta = \beta_0$ against $\beta \neq \beta_0$. Note that this testing situation includes the important special case where $\beta_0 = 0$. When β_0 is indeed zero, recording the value of a covariable X is useless effort with respect to enlightening the experimenter with regard to the principal variable Y. When $H_0: \beta = \beta_0$ is true, then

$$\frac{(b-\beta_0)\sqrt{\sum\limits_{i=1}^{n} x^2_i}}{\sigma} \sim N(0, 1)$$

and

$$\frac{(b - \beta_0) \sqrt{\sum_{i=1}^{n} x^2_i}}{\hat\sigma} \sim t(n - 2).$$

To effect a test of H_0 when σ is unknown, we compare the calculated value

$$\frac{(b - \beta_0) \sqrt{\sum_{i=1}^{n} x^2_i}}{\hat\sigma}$$

with the tabulated Student $t(n - 2)$ critical values at the desired level of the probability of a Type I error.

The special case of testing $\beta = 0$ can be effected directly from the regression AOV. Recalling that a two-tailed t test is the same as a one-tailed (upper) F test and that $t^2(n - 2) = F(1, n - 2)$, it is seen that the ratio of mean square $R(\beta|\mu)$ to the mean square for residual when compared to the proper $F(1, n - 2)$ critical value preforms the same test as the t test described above.

The distribution properties of b and $\hat\sigma^2$ immediately yield confidence interval statements for β and σ^2. These statements are for β in the case where σ^2 is unknown, and for σ^2

$$P\left[b - t \frac{\hat\sigma}{\sqrt{\sum_{i=1}^{n} x^2_i}} \le \beta \le b + t \frac{\hat\sigma}{\sqrt{\sum_{i=1}^{n} x^2_i}} \right] = 1 - \alpha$$

$$P\left[\frac{\hat\sigma^2}{d} \le \sigma^2 < \frac{\hat\sigma^2}{c} \right] = 1 - \alpha$$

where t, c, and d are the proper tabulated Student $t(n - 2)$ and $\chi^2(n - 2)$ values.

5.8 AN ACADEMIC ILLUSTRATION

Consider now a purely academic problem in which all parameters for a linear regression model are known. Observations first will be created then later analyzed in order to study and compare estimates with known parameters. With these objectives in mind, consider the (true) population regression line $\mu_Y = .4 + .5X$. This of course means that we have chosen $\alpha = .4$ and $\beta = .5$. Observations Y_i are now created by letting

$Y_i = .4 + .5X_i + \epsilon_i$ where $\epsilon_i \sim \text{NID}(0, \sigma^2)$, σ^2 is conveniently chosen to be .5 and the X_i are chosen as follows:

Choose X	and read from a random normal table
$X_1 = 1$	$\epsilon_1 = \quad .2$
$X_2 = 1$	$\epsilon_2 = -.8$
$X_3 = 2$	$\epsilon_3 = -.1$
$X_4 = 6$	$\epsilon_4 = -.3$
$X_5 = 7$	$\epsilon_5 = \quad .4$
$X_6 = 7$	$\epsilon_6 = 1.2$

Note that the sample variance of the ϵ_i is .464 which agrees with the chosen value of $\sigma^2 = .5$. The created observations augmented by some calculations are:

	X	x	xY	x^2
$Y_1 = .4 + .5 + .2 = \quad 1.1$	1	-3	-3.3	9
$Y_2 = .4 + .5 - .8 = \quad .1$	1	-3	$-\ .3$	9
$Y_3 = .4 + 1.0 - .1 = \quad 1.3$	2	-2	-2.6	4
$Y_4 = .4 + 3.0 - .3 = \quad 3.1$	6	2	6.2	4
$Y_5 = .4 + 3.5 + .4 = \quad 4.3$	7	3	12.9	9
$Y_6 = .4 + 3.5 + 1.2 = 5.1$	7	3	15.3	9
Totals	24	0	28.2	44

Having chosen the X_i so that $\bar{X} = 4$, the value of μ is $\mu = \alpha + \beta\bar{X} = .4 + .5(4) = 2.4$ and the model can be written in terms of x_i as $Y_i = 2.4 + .5x_i + \epsilon_i$. The regression AOV for the six created observations is:

Source	df	SS	MS
Total	6	57.02	
$R(\mu)$	1	37.50	
$R(\beta|\mu)$	1	18.07	18.07
Residual	4	1.45	.3625

Had our objective been to test $H_0: \beta = 0$, based on this data, we would have rejected emphatically the null hypothesis.

Displayed in Table 5.2 are the selected true parameters along with estimates obtained from the six observations.

In Figure 5.8 the selected population regression line is pictured along with the created observations and the least squares line computed from the observations. Notice that the least squares line fits the data better than the population regression line.

Table 5.2   Parameters and Their Estimates

Selected Values of the Parameter	Estimate
$\alpha = .4$	$a = - .0636$
$\beta = .5$	$b = .6409$
$\mu = 2.4$	$\bar{Y} = 2.5$
$\sigma^2 = .5$	$\hat{\sigma}^2 = .3625$

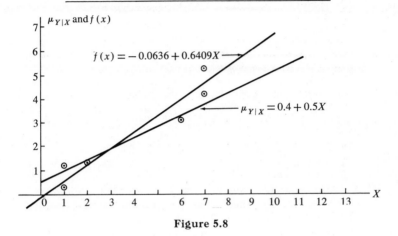

Figure 5.8

5.9   EXERCISES

(1)   Create observations $Y_1, \ldots, Y_6$ for the situation described in the illustrated example of Section 5.8 modified by fixing X values at $X_1, \ldots, X_6$ equal to 3, 3, 4, 4, 5, and 5, respectively.   For comparison purposes, suppose that the ϵ_i are the same as those in Section 5.8. From the observations, compute estimates of α, β, μ, and σ^2. Write out the regression AOV and note that the F ratio $R(\beta|\mu)$/residual mean square is now considerably different from the corresponding F ratio in the case where $X_1, \ldots, X_6$ were 1, 1, 2, 6, 7, and 7, respectively. Explain the difference in the values of the F ratio.

(2)   For the notation of this chapter show that

$$\sum_{i=1}^{n} \bar{Y}bx_i = \sum_{i=1}^{n} \bar{Y} d_i = \sum_{i=1}^{n} bx_i d_i = 0.$$

(3)   Work out the details of showing that $\mathcal{E}(b) = \beta$ and $\text{Var}(b) = \sigma^2 \bigg/ \sum_{i=1}^{n} x^2_i$ when a simple linear regression model is assumed and the X_i are fixed.

(4)   With $\alpha = .01$, use the following data to test $\beta = 0$ against $\beta \neq 0$ by writing out the regression AOV.

$X:$ 8   8   7   7   1   1   0   0
$Y:$ 2.7   2.6   2.7   2.3   2.0   2.5   2.4   2.3

(5)   Estimate σ^2 from the data in Exercise (4).

(6)   Write out regression AOV's and set 95 percent confidence intervals on the values of β and σ^2 for each of the following problems in Section 5.5: (1), (2), (5), and (6).

(7)   Consider again the data in Exercise (4) but now let $Z_i = X_i - 2$ and write out the regression AOV for the regression of Y on Z. Compare the AOV with that of Exercise (4) and study the geometry of the situation.

(8)   Assume, for the following data, that the simple linear regression models $Y_i = \alpha_1 + \beta_1 X_i + \epsilon_i$ and $Z_j = \alpha_2 + \beta_2 W_j + \delta_j$ where $\epsilon_i \sim NID(0, \sigma^2)$, $\delta_j \sim NID(0, \sigma^2)$ and each ϵ_i is independent of each δ_j.

$X:$ 22   31   25   29   24 $W:$ 40   35   33   38   37   35
$Y:$ 75   79   80   81   76 $Z:$ 70   68   67   68   66   71

(a)   Determine the least squares regression line for the regression of Y on X.
(b)   Determine the least squares regression line for the regression of Z on W.
(c)   With $\alpha = .05$, test the null hypothesis that the two regressions have common slope. [*Hint*: Determine first the distribution of $(b_1 - b_2)$.]

(9)   The following data are a set of yields of strawberries in pounds per plot.

Experimental plot
number: 1   2   3   4   5   6   7   8
Yield Y: 9.75   6.38   7.93   5.92   6.50   8.10   8.24   7.54

The model $Y = \mu + \epsilon$ is assumed where $\epsilon \sim N(0, \sigma^2_1)$.
(a)   Write out an AOV.
(b)   Estimate the mean yield μ and the experimental error σ^2_1.

The amount of nitrogen for the above experimental plots was

Experimental plot number: 1   2   3   4   5   6   7   8
X in pounds: 20   5   15   5   10   20   15   10

Now the model is $Y = \mu + \beta x + \epsilon$ where $x = X - \bar{x}$ and $\epsilon \sim N(0, \sigma^2_2)$.
(c)   Write out the AOV for the regression model.
(d)   Estimate the experimental error σ^2_2.
(e)   Determine the linear regression line $Y = a + bX$.
(f)   Plot the observations, draw the regression line, and estimate the mean yield if 18 pounds of nitrogen had been added to several experimental plots.

5.10   THE CASE WHERE A LINEAR REGRESSION MODEL MAY NOT FIT

The choice of a linear model in experimentation is usually backed up by theory, experience, or observation. There does exist, however, a statistical procedure to help decide whether or not a linear model is applicable for a prescribed range of the auxiliary variable. We recall first that experimental error is the name given to variation among experimental units that have been treated alike. In the context of regression where only one auxiliary variable is deemed worth studying, two experimental units have been treated alike if they possess the same numerical value for the auxiliary variable. Thus, when values of an auxiliary variable can be chosen in advance of running an experiment by repeating values of the auxiliary variable, an estimate of the experimental error variance can be obtained independent of whether or not a linear regression model fits the data. A measure of the *lack of fit* of the model is the difference between the mean square for residual and the mean square for experimental error divided by the residual mean square.

The concept of a mean square measure for lack of fit of the model will be elaborated upon by referring to an illustrative example. Consider the data:

$$X:\ 0\quad 1\quad 1\quad 2\quad 2\quad 2\quad 3\quad 3\quad 4$$
$$Y:\ 9.0\quad 7.2\quad 6.4\quad 3.9\quad 3.8\quad 4.3\quad 2.8\quad 3.2\quad 2.6.$$

The observations are pictured in Figure 5.9 and a regression AOV for the data follows the figure.

Source	df	SS	MS	$\mathcal{E}MS$	F
Total	9	246.78			
$R(\mu)$	1	207.36			
$R(\beta\|\mu)$	1	34.68			
Experimental error and lack of fit	3	4.20	1.400	$\sigma^2 + L$	
Experimental error only	4	.54	.135	σ^2	10.37

The sum of squares for total, $R(\mu)$ and $R(\beta|\mu)$ are computed in the conventional manner as illustrated earlier in the chapter. Sum of squares for experimental error only, is the pooled within sum of squares associated with the X values 1, 2, and 3, and of course the degrees of freedom are the pooled degrees of freedom, for the observations at these values of X. The sum of squares and the degrees of freedom for experimental error

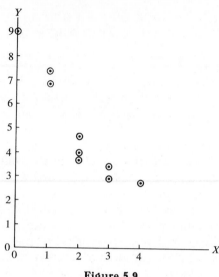

Figure 5.9

and lack of fit is obtained by subtraction. In this illustration the F ratio for the last two sources of variation is 10.37, indicating that lack of fit is present in significant magnitude. The tabulated $F(3, 4)$ value is 6.59 when a Type I error of .05 is used. When lack of fit is not present, the mean squares, for each of the last two entries in the table, are estimating σ^2 which then implies that the F ratio in that case should be as large as 6.59 in only 5 percent of the time.

5.11   ADJUSTED SCORES

One very useful application of the theory of regression is the adjustment of responses in light of known differences in characteristics of the experimental units. To illustrate the application, consider the following data recorded in connection with strength tests for school boys.

Weight of the boy (X):   100   100   120   120   140   140
Pounds lifted (Y):   95   85   90   95   105   100

The least squares line for the regression of Y on X is

$$f(X) = 95 + \frac{5}{16} x = 57.5 + \frac{5}{16} X.$$

Tabulations recorded in Table 5.3 illustrate the concept of *adjusted scores*.

TABLE 5.3

X	Y	$f(X)$	$d = Y - f(X)$	$\bar{Y} + d$
100	95	88.75	6.25	101.25
100	85	88.75	−3.75	91.25
120	90	95	−5.00	90.00
120	95	95	0.00	95.00
140	105	101.25	3.75	98.75
140	100	101.25	−1.25	93.75

To focus the discussion, consider the first boy. The difference between Y_1 and $f(100)$ is 6.25. It might be said that this boy lifted 6.25 pounds more than was expected. Since the average weight lifted was 95 pounds and since he lifted 6.25 more than expected then his score adjusted for his weight is placed at 101.25 pounds. Notice that this adjusted score is greater than the adjusted score for the 140-pound boy who lifted 105 pounds.

5.12 EXERCISES

(1) Determine the mean squares for experimental error only, and experimental error plus lack of fit when a simple linear regression model is assigned to the data:

X: 0 1 1 2
Y: 3 0 2 0

(2) Test the goodness of fit of a simple linear regression model to the data:

X: 1   1   2   2   2   4
Y: 7.0 5.4 3.0 3.2 4.6 2.6

(3) Consider the quadratic equation $Y = X^2 - 6X + 8$. Determine a least squares line for the points

X: 1   7   8   9
Y: 3 15 24 35

all of which lie on the quadratic curve. Plot these points, draw the least squares line and sketch the quadratic curve $Y = X^2 - 6X + 8$. Would you say that the linear fit of the data was good?

(4) Examine the data in Exercise (4) of Section 5.9 with goodness of fit of the model the main object of concern.

(5) Adjust the following Y scores for the variability in the X scores.

Y: 25 29 38 40
X: 4   5   6   7

(6) Adjust the following Y scores for the variability in the X scores.

Y: 79 52 67 49 53
X: 10 16 15 20 14

(7) A laboratory-type problem: A farmer-researcher for ten years kept careful records of corn production and related information for his northwest forty acres. As might be expected, there was a strong relationship between yield and rainfall. In the following data, adjust the yields Y for the rainfall in May, X.

Year:	1951	1952	1953	1954	1955	1956	1957	1958	1959	1960
Y:	11.2	11.4	5.1	5.0	11.5	12.7	14.9	13.8	13.9	12.2
X:	2.1	3.7	1.9	1.8	2.5	3.8	5.2	7.1	6.0	2.9

If in May 1961 the rainfall was 2.0, what could he expect his corn yield to be for his northwest forty acres?

6

STATISTICAL INFERENCE WHEN INTERACTION IS PRESENT

6.1 THE CONCEPT OF INTERACTION

Many new problems in life in general but more specifically in experimental sciences are merely old problems in a new context. In this chapter some experimental situations will be described and then, whenever appropriate, attention will be called to the fact that the new problem is merely another form of a problem already considered or a generalization of a previous problem situation. The approach to a new problem and the resulting methods advocated for a solution will often be seen to be refinements of methods already studied. Major objectives of the chapter are to give more insight into the concepts presented earlier and to familiarize the student with some of the scientific language in use today so that he will recognize in a presented situation the structure of the data and thereby be more able to use the proper techniques for analyzing the data. In this chapter much emphasis will be given to statistical models and experiments with factorial arrangements of treatments.

Consider the following situation, which shows that the model $Y_{ij} = \mu + \beta_i + \gamma_j + \epsilon_{ij}$ may not always describe two-way cross-classification data adequately. At supermarkets in the spring of the year, one can buy dry dirt packages wrapped in cellophane paper. In the dirt are seeds and on the package can be found words to the effect that, after removing the paper, applying water, and setting in a sunny place, petunia plants will appear. Suppose that four of these packages were purchased with the following treatment combinations randomly assigned to the packages.

$WS \equiv$ water applied and the package placed in the sunlight.
$\bar{W}S \equiv$ no water applied but the package placed in the sunlight.
$W\bar{S} \equiv$ water applied but the package placed in the dark.
$\overline{WS} \equiv$ no water applied and the package placed in the dark.

Suppose that the following amounts of green matter appeared after a fixed time had elapsed.

					Totals
	WS	9.8	$\bar{W}S$	.3	10.1
	$W\bar{S}$	.8	$\overline{WS}$	.1	.9
Totals		10.6		.4	11.0

In this experiment, we have strong indications that the effect of water is much different in the presence of sunlight from what it is in the presence of no sunlight. The same can be said for the effect of sunlight with respect to water. A change from no sunlight to sunlight, in the absence of water, increased the response .2 units. A change from no water to water in the absence of sunlight increased the response .7 units. The linear additive effect, ignoring experimental error, would give an increased response of 1 unit. The observed response of 9.8 for the treatment combination WS indicates that sunlight and water interact to produce growth of magnitude equal to 8.8 units above that of the combined linear effects. Notice that the *interaction effect* of 8.8 units can be written as the sum of the responses for WS and $\overline{WS}$ minus the sum of the responses for $W\bar{S}$ and $\bar{W}S$. This combination of responses is a special case of the concept of contrasts which will be studied in this chapter.

6.2 FACTORS AND CONTRASTS

Some experimenters like to refer to sources of variability as factors in an experiment. Thus, in a randomized block experiment involving several varieties, one factor is the varieties and another is the blocks. Although the word factor is used in conjunction with qualitative sources of variability such as varieties, it is most frequently used in experiments where classification of populations is done on a quantitative scale and the experimenter arbitrarily, for each experimental unit, sets each index of classification at a fixed value. To illustrate, consider a cake-baking experiment where various oven times and oven temperatures are suggested. These are quantitatively indexed factors. Selecting, for example, three oven temperatures and four oven times, the experiment could be described as a two-factor experiment, one factor at three levels and the other at four levels. The experiment could also be described as one involving a factorial arrangement of 12 treatment combinations. In general, if there are l_i levels of the ith factor and f factors, then there are $l_1 l_2 \cdots l_f$ possible treatment combinations. To be considered in the next sections of this chapter is the case where there are two factors both at two levels. This situation is often denoted as a 2^2 factorial experiment.

Definitions and results in the next discussion will be framed in the context of experiments where equal numbers of observations have been taken from each of the populations under consideration. This statement, when applied to factorial experiments, means that each possible treatment combination has been repeated the same number of times. Let T_α, $\alpha = 1, 2, \ldots, t$ be observations or observation totals for the t different populations. A sum $C = \sum_{\alpha=1}^{t} k_\alpha T_\alpha$ is called a *linear combination* of the observations or totals T_α. If the constants k_α have the property that $\sum_{\alpha=1}^{t} k_\alpha = 0$, the linear combination is called a *contrast*. When all but two of the coefficients are zero, a contrast is called a *simple contrast*. Two contrasts $C_1 = \sum_{\alpha=1}^{t} k_{1\alpha} T_\alpha$ and $C_2 = \sum_{\alpha=1}^{t} k_{2\alpha} T_\alpha$ of the same observations or totals are called *orthogonal* if $\sum_{\alpha=1}^{t} k_{1\alpha} k_{2\alpha} = 0$.

$\bar{Y} = (1/n)Y_1 + (1/n)Y_2 + \cdots + (1/n)Y_n$ is an example of a linear combination.

$C_1 = T_1 + T_2 - T_3 - T_4$ is an example of a contrast.

$C_2 = T_1 - T_2$ is an example of a simple contrast.

$C_1 = T_1 + T_2 - T_3 - T_4$ and $C_2 = T_1 - T_2$ are orthogonal contrasts.

6.3   STATISTICAL MODELS WHEN INTERACTION IS PRESENT

Having briefly introduced the concept of treatment combinations, interaction, contrasts, and orthogonality, the objective now is to relate these ideas to earlier concepts and in particular to tie them to analysis of variance. Suppose that the following treatment combinations were applied to similar pans of homogeneous cake batter.

Designation of the treatment combination	Oven temperature in degrees	Oven time in minutes
(1, 1)	350	40
(1, 2)	350	50
(2, 1)	400	40
(2, 2)	400	50

Consider first the case where each treatment combination is applied to one pan of batter and the texture response for the (i, j) treatment combination is denoted by Y_{ij}. Three important orthogonal contrasts of these observations are:

$B = Y_{11} + Y_{12} - Y_{21} - Y_{22}$ which compares low with high levels of temperature and will be called the temperature contrast.

$C = Y_{11} - Y_{12} + Y_{21} - Y_{22}$ which compares low with high levels of time and will be called the time contrast.

$I = Y_{11} - Y_{12} - Y_{21} + Y_{22}$ which compares the low-low and high-high with low-high and high-low and has already been called the interaction contrast.

The interaction ideas to be studied perhaps can best be understood by considering statistical models that adequately describe data when interaction is present. If the temperature effect on texture is denoted by β_i and the time effect on texture is denoted by γ_j, then the model $Y_{ij} = \mu + \beta_i + \gamma_j + \epsilon_{ij}$ can describe interaction data only if the interaction effect is included in the component ϵ_{ij}, in which case it is wrong to assume that $\epsilon_{ij} \sim \text{NID}(0, \sigma^2)$ and as a matter of fact, when treatment combinations are not repeated, there is no measure of experimental error. Recall the important definition that experimental error variance is a measure of the variability of experimental units that have been treated alike.

If now each of the 2^2 treatment combinations is repeated r times, consideration of the experiment and the data might lead us to adopt for the texture responses the model $Y_{ijk} = \mu + \beta_i + \gamma_j + \delta_{ij} + \epsilon_{ijk}$ where $k = 1, 2, \ldots, r$, δ_{ij} is the interaction effect of the treatment combination (i, j) and the random errors $\epsilon_{ijk} \sim \text{NID}(0, \sigma^2)$. Counting the parameters involved, we find that there are ten of them. This presents difficulties in estimation of parameter values and testing hypotheses concerning the parameters. To help understand the difficulties present in this situation, consideration will be given to a similar but simpler situation studied earlier. In the two-population fixed-effects model $Y_{ij} = \mu + \tau_i + \epsilon_{ij}, i = 1, 2,$ $j = 1, 2, \ldots, r$, there are four parameters. Infinitely many sets of values of $\mu, \tau_1,$ and τ_2 can be chosen to give the same mean values of observations Y_{1j} and Y_{2j}. Essentially the same thing can be said in another way. Although unbiased estimates of $(\mu + \tau_1)$, $(\mu + \tau_2)$, and $(\tau_1 - \tau_2)$ exist, there exists no function of the observations which is an unbiased estimator of τ_1, or τ_2, or μ alone. The parameters are said to be confounded and the model is said to overparameterize the experimental situation. The restriction that $\tau_1 + \tau_2 = 0$ removes most of the estimation and hypotheses-testing difficulties with $Y../2r$ and $Y_{i.}/r - Y../2r$ now serving as unbiased estimators of μ and τ_i. Restrictions other than $\tau_1 = -\tau_2$ could be used to remove the overparameterization, but this is the natural most-often-used restriction.

The model $Y_{ijk} = \mu + \beta_i + \gamma_j + \delta_{ij} + \epsilon_{ijk}$ overparameterizes the interaction, cake-baking, experimental situation. Examining the expected value of the temperature contrast of treatment combination totals, we obtain

$$\mathcal{E}[Y_{11.} + Y_{12.} - Y_{21.} - Y_{22.}] = 2r(\beta_1 - \beta_2) + r(\delta_{11} + \delta_{12} - \delta_{21} - \delta_{22}).$$

Likewise

$$\mathcal{E}[Y_{11.} - Y_{12.} + Y_{21.} - Y_{22.}] = 2r(\gamma_1 - \gamma_2) + r(\delta_{11} - \delta_{12} + \delta_{21} - \delta_{22})$$

and

$$\mathcal{E}[Y_{11.} - Y_{12.} - Y_{21.} + Y_{22.}] = r(\delta_{11} - \delta_{12} - \delta_{21} + \delta_{22}).$$

The best restriction for resolving the difficulties of our parameterization seems to be to require that $\beta_1 + \beta_2 = 0$, $\gamma_1 + \gamma_2 = 0$, $\delta_{11} + \delta_{12} = 0$, $\delta_{21} + \delta_{22} = 0$, and $\delta_{11} + \delta_{21} = 0$. These restrictions in effect leave us with one temperature parameter, one time parameter, and one interaction of time and temperature parameter. Applying these restrictions, the expected values of the above contrasts become: $\mathcal{E}(B) = 4r\beta_1$, $\mathcal{E}(C) = 4r\gamma_1$, $\mathcal{E}(I) = 4r\delta_{11}$. Notice that $4r$ is the total number n of observations so that $\mathcal{E}[(1/n)B] = \beta_1 = -\beta_2, \mathcal{E}[(1/n)C] = \gamma_1 = -\gamma_2$ and $\mathcal{E}[(1/n)I] = \delta_{11} = -\delta_{12} = -\delta_{21} = \delta_{22}$. Hence these restrictions allow us to estimate the five independent parameters left after making the restriction and, as will soon be seen, they allow for testing of hypotheses relative to the interaction data.

6.4   TESTING HYPOTHESES AND THE AOV
FOR 2^2 FACTORIAL EXPERIMENTS

In this section, we shall consider first the AOV for the temperature time 2^2 factorial situation when r, the number of replicates of each treatment combination, is 1. Using the fixed-effects model $Y_{ij} = \mu + \beta_i + \gamma_j + \eta_{ij}$

to describe the data, part of the AOV is as follows:

Source	df	SS
Total	4	$\sum_i \sum_j Y^2_{ij}$
Mean	1	$Y^2_{..}/4$
Temperature (block)	1	$\frac{1}{2} \sum_i Y^2_{i.} - (Y^2_{..}/4)$
Time (treatment)	1	$\frac{1}{2} \sum_i Y^2_{.j} - (Y^2_{..}/4)$
Residual	1	$\sum_i \sum_j Y^2_{ij} - \frac{1}{2} \sum_i Y^2_{i.} - \frac{1}{2} \sum_j Y^2_{.j} + (Y^2_{..}/4)$

The model and the AOV are seen to be similar to those in a randomized block experiment where temperatures play the role of blocks and times play the role of treatments. The last source of variability is referred to as residual rather than experimental error because, when $r = 1$, interaction is present and there are no experimental units treated alike and there exists no measure of experimental error. The word "treated" here is used in the sense that no treatment combination of temperature and time was duplicated. The symbols η_{ij} in the model can be thought of as representing interaction effects plus random-error effects, that is, $\eta_{ij} = \delta_{ij} + \epsilon_{ij}$. *It should be pointed out that in Chapter 4 the interaction effects of blocks with treatments were tacitly assumed to be zero, thus justifying the use there of the term experimental error.*

The temperature sum of squares expressed in the AOV can be written in the following ways.

$$\frac{1}{2} \sum_i Y^2_{i.} - \frac{Y^2_{..}}{4} = \frac{1}{4} \left[2 \sum_i Y^2_{i.} - (Y_{1.} + Y_{2.})^2 \right]$$

$$= \frac{1}{4}[Y^2_{1.} - 2Y_{1.}Y_{2.} + Y^2_{2.}]$$

$$= \frac{1}{4}[Y_{1.} - Y_{2.}]^2 = \frac{1}{4}[Y_{11} + Y_{12} - Y_{21} - Y_{22}]^2.$$

In the last form the temperature sum of squares is exhibited as a function of the temperature contrast B. In a similar way it can be shown that the time sum of squares $= C^2/4$ where C is the time contrast $Y_{11} - Y_{12} + Y_{21} - Y_{22}$, and the residual sum of squares $= I^2/4$, where I is the interaction contrast $Y_{11} - Y_{12} - Y_{21} + Y_{22}$. Although it will not be proved, the fact that the B, C, and I contrasts are orthogonal implies that the corresponding sums of squares are independent random variables. As a last comment before the case where $r > 1$ is considered, reference is made to the use of interaction as a source of variability instead of residual because of the fact that the sum of squares is $I^2/4$.

Consider now the cake-baking experiment when each treatment combination is applied to r cakes. The kth response to the (i, j) treatment combination will be denoted by Y_{ijk} and the fixed-effects model $Y_{ijk} = \mu + \beta_i + \gamma_j + \delta_{ij} + \epsilon_{ijk}$ will be assumed for the $4r$ observations. In order to test hypotheses and eliminate overparameterization, the restrictions $\beta_1 = -\beta_2, \gamma_1 = -\gamma_2, \delta_{11} = -\delta_{12} = -\delta_{21} = \delta_{22}$ will be invoked. As usual, it is assumed that $\epsilon_{ijk} \sim \text{NID}(0, \sigma^2)$. The data, contrast coefficients, and AOV appear as:

DATA

Treatment combinations	(1, 1)	(1, 2)	(2, 1)	(2, 2)
	Y_{111}	Y_{121}	Y_{211}	Y_{221}
	.	.	.	.
Observations	.	.	.	.
	.	.	.	.
	Y_{11r}	Y_{12r}	Y_{21r}	Y_{22r}
Totals	$Y_{11.}$	$Y_{12.}$	$Y_{21.}$	$Y_{22.}$
Contrast coefficients k_{ij} for B:	1	1	-1	-1
Contrast coefficients k_{ij} for C:	1	-1	1	-1
Contrast coefficients k_{ij} for I:	1	-1	-1	1

AOV

Source	df	SS	MS	$\mathcal{E}MS$
Total	$4r$	$\displaystyle\sum_k\sum_j\sum_i Y^2_{ijk}$		
Mean	1	$\dfrac{Y^2_{...}}{4r}$		
Temperature	1	$\dfrac{1}{2r}\displaystyle\sum_i Y^2_{i..} - \dfrac{Y^2_{...}}{4r}$	$\dfrac{1}{4r}B^2$	$r(\beta_1 - \beta_2)^2 + \sigma^2$
Time	1	$\dfrac{1}{2r}\displaystyle\sum_j Y^2_{.j.} - \dfrac{Y^2_{...}}{4r}$	$\dfrac{1}{4r}C^2$	$r(\gamma_1 - \gamma_2)^2 + \sigma^2$
Interaction	1	$\displaystyle\sum_i\sum_j \dfrac{(Y^2_{ij.})}{r} - \dfrac{1}{2r}\displaystyle\sum_i Y^2_{i..}$ $-\dfrac{1}{2r}\displaystyle\sum_j Y^2_{.j.} + \dfrac{Y^2_{...}}{4r}$	$\dfrac{1}{4r}I^2$	$r\displaystyle\sum_{ij}\delta^2_{ij} + \sigma^2$
Experimental error	$4(r-1)$	$\displaystyle\sum_k\sum_i\sum_j \left(Y_{ijk} - \dfrac{Y_{ij.}}{r}\right)^2$	MSE	σ^2

The sum of squares in the AOV can be obtained either by the formulas or by applying the proper coefficients in the contrast expression $\sum_i \sum_j k_{ij} Y_{ij\cdot}$. The algebra of showing that [contrast]$^2/n$ equals the sum of squares associated with a particular source is left as an exercise.

The AOV is especially useful in testing hypotheses concerned with contrasts and a principal reason for introducing statistical models involving interaction was to aid in the discussion of hypothesis testing. With the model and parameter restrictions of this section, the expected mean squares are those set forth in the last column of the AOV. If it is desired to test a hypothesis that temperature or time or interaction effects are zero, then one of the several ways of expressing these hypotheses is $\sum_i \beta^2_i = 0$, $\sum_i \gamma^2_i = 0$, and $\sum_i \sum_j \delta^2_{ij} = 0$. As seen from the AOV, when the hypothesis under question is indeed true, the expected mean square for the source of variation associated with the particular hypothesis estimates σ^2. The mean squares are under normal theory independent variables and each of the ratios $B^2/4r$ MSE, $C^2/4r$ MSE, and $I^2/4r$ MSE are Snedecor F variables when the corresponding hypothesis is true. A test of a null hypothesis can then be effected by comparing the appropriate ratio of mean squares with the proper tabulated F value.

Before leaving the subject of expected mean squares for 2^2 factorial experiments, mention is made again of the fact that only the fixed-effects model was considered in describing the interaction data. In a later chapter coverage will be given to the matter of testing and estimation in 2^2 factorial experiments when a linear model is assumed in which some or all of the effects are random.

6.5 EXERCISES

(1) Show that if $B =$ temperature contrast discussed in the example of Section 6.2, then $B^2/n =$ sum of squares associated with the temperature factor.

(2) Determine the number of observations in the following experimental situations and write out a notation for each observation.
  - (a) 2 factors each at 2 levels with each treatment combination replicated 3 times.
  - (b) 3 factors each at 2 levels with each treatment combination replicated 4 times.
  - (c) 2 factors each at 2 levels with each treatment combination replicated 2 times.
  - (d) 3 factors each at 3 levels with each treatment combination replicated 2 times.

(e) 4 factors each at 2 levels with each treatment combination replicated 3 times.

(3) From the following linear combinations, pick out the contrasts and then determine subsets of the contrasts which contain orthogonal contrasts.

$$C_1 = Y_1 + Y_2 + Y_3 + Y_4$$
$$C_2 = Y_1 - 2Y_2 + Y_3$$
$$C_3 = Y_1 - 2Y_2 + 3Y_3 - 2Y_4$$
$$C_4 = 3Y_1 - 2Y_2 - 2Y_3 + Y_4$$
$$C_5 = Y_1 + Y_2 + Y_3 - 3Y_4$$

(4) Consider an ice-cream texture experiment involving the factors, sugar content and fat content, each at two levels. Denote the ith level of sugar and the jth level of fat by the treatment combination symbol (i, j) and suppose that each treatment combination is replicated giving the following set of responses:

Treatment combination	(1, 1)	(1, 2)	(2, 1)	(2, 2)
	2.2	5.8	4.7	6.0
	2.8	5.6	4.5	5.6

(a) Write out the contrasts for sugar, fat, and the interaction of sugar and fat.
(b) Write out the AOV for this 2^2 factorial experiment.
(c) Write out a fixed-effects model to describe the kth observation for treatment combination (i, j).
(d) With $\alpha = .05$ use the AOV to make significance tests of the following null hypotheses assuming, of course, that your fixed-effects model describes the data:
   (1) The levels of sugar produce the same effects when the fat level is held fixed.
   (2) The levels of fat produce the same effects when the sugar level is held fixed.
   (3) The interaction of sugar and fat is negligible.

(5) Consider factors $\mathcal{A}$ and $\mathcal{B}$ both at low and high levels. To help understand the concept of interaction, compute the interaction contrast for each situation and picture for the last four situations the responses in a manner similar to that done here for the first situation (see Figure 6.1).

(6) Assume the fixed-effects model $Y_{ijk} = \mu + \beta_i + \gamma_j + \delta_{ij} + \epsilon_{ijk}$, $i = 1, 2$; $j = 1, 2$; $k = 1, \ldots, r$; where $\epsilon_{ijk} \sim NID(0, \sigma^2)$ and let the restrictions $\beta_1 + \beta_2 = 0$, $\gamma_1 + \gamma_2 = 0$, and $\delta_{11} = \delta_{22} = -\delta_{12} = -\delta_{21}$ be applied. Work out the expected values for the main-effects contrasts, for the interaction contrast, and for the mean squares in the AOV corresponding to the main effects, interaction, and experimental error.

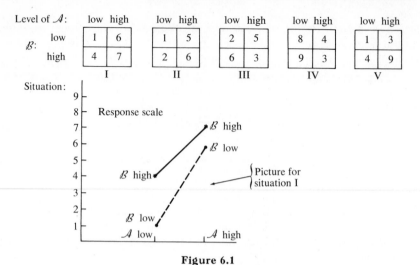

Figure 6.1

(7) Let observations Y_{ijk} be given by the linear fixed-effects model $Y_{ijk} = \mu + \beta_i + \gamma_i + \delta_{ij} + \epsilon_{ijk}$ where $\epsilon_{ijk} \sim \text{NID}(0, \sigma^2)$ and the parameters involved have been chosen to be $\mu = 8$, $\beta_1 = -1$, $\beta_2 = 1$, $\gamma_1 = 3$, $\gamma_2 = -3$, $\delta_{11} = \delta_{22} = -2$, $\delta_{12} = \delta_{21} = 2$, and $\sigma^2 = 4$. Create the twelve observations corresponding to the random errors

$\epsilon_{111} = -2.8$	$\epsilon_{121} = -.7$	$\epsilon_{211} = 2.7$	$\epsilon_{221} = 3.3$
$\epsilon_{112} = 1.8$	$\epsilon_{122} = -.1$	$\epsilon_{212} = 2.0$	$\epsilon_{222} = -1.0$
$\epsilon_{113} = -3.0$	$\epsilon_{123} = -1.3$	$\epsilon_{213} = -.3$	$\epsilon_{223} = -.2$

Analyze the created observations with unbiased estimation of the five functionally independent parameters as the main objective. Compare estimates with the given values of the parameters.

6.6 2^3 FACTORIAL EXPERIMENTS

Let three factors α, $\mathcal{B}$, and $\mathcal{C}$ be investigated in an experiment where each factor is at two levels and the eight treatment combinations are replicated r times. Let (i, j, k) denote the treatment combination consisting of the ith level of α, the jth level of $\mathcal{B}$, and the kth level of $\mathcal{C}$. The eight treatment combinations are then designated by $(1, 1, 1)$; $(1, 1, 2)$; $(1, 2, 1)$; $(1, 2, 2)$; $(2, 1, 1)$; $(2, 1, 2)$; $(2, 2, 1)$; $(2, 2, 2)$. Such a situation is called a 2^3 factorial experiment with r replications. The response for the lth replicate of treatment combinations (i, j, k) will be denoted by Y_{ijkl} and there will be $r(2^3)$ responses. With three factors there are four interactions to consider. The interactions will be denoted by $\alpha\mathcal{B}$, $\alpha\mathcal{C}$,

$\mathcal{BC}$, and $\mathcal{ABC}$ with the last interaction referred to as the second-order interaction of the experiment. Contrasts corresponding to the above factors and interactions will be denoted by the capital letters A, B, C, I_{AB}, I_{AC}, I_{BC}, and I_{ABC}, each contrast being associated with one degree of freedom. An AOV for such an experiment might appear as follows:

Source	df	MS	$\mathcal{E}MS$ for the fixed model with restrictions on the parameters
Total	$8r$	$\displaystyle\sum_i \sum_j \sum_k \sum_l Y^2_{ijkl}/8r$	
Mean	1	$(Y\ldots)^2/8r$	
$\mathcal{A}$	1	$A^2/8r$	$\sigma^2 + 4r(\alpha^2_1 + \alpha^2_2)$
$\mathcal{B}$	1	$B^2/8r$	$\sigma^2 + 4r(\beta^2_1 + \beta^2_2)$
$\mathcal{C}$	1	$C^2/8r$	$\sigma^2 + 4r(\gamma^2_1 + \gamma^2_2)$
$\mathcal{AB}$	1	$(I_{AB})^2/8r$	$\sigma^2 + 2r \displaystyle\sum_i \sum_j (\alpha\beta)^2_{ij}$
$\mathcal{AC}$	1	$(I_{AC})^2/8r$	$\sigma^2 + 2r \displaystyle\sum_i \sum_k (\alpha\gamma)^2_{ij}$
$\mathcal{BC}$	1	$(I_{BC})^2/8r$	$\sigma^2 + 2r \displaystyle\sum_i \sum_k (\beta\gamma)^2_{jk}$
$\mathcal{ABC}$	1	$(I_{ABC})^2/8r$	$\sigma^2 + r \displaystyle\sum_i \sum_j \sum_k (\alpha\beta\gamma)^2_{ijk}$
Error	$8(r-1)$	$\dfrac{\text{Pooled within } SS}{8(r-1)}$	σ^2

Perhaps the easiest way to calculate the desired contrasts is to form a table of contrast coefficients such as that exhibited here in Table 6.1.
The contrast is then obtained by applying the proper coefficients to the totals and adding. The reader can check that the seven contrasts are orthogonal. For a specific factor, the main effects are merely the low level of the factor versus the high level. For the interaction of two factors, say $\mathcal{A}$ and $\mathcal{B}$, the contrast is one of high for both plus low for both versus low for $\mathcal{A}$ and high for $\mathcal{B}$ plus high for $\mathcal{A}$ and low for $\mathcal{B}$. The coefficients for a first-order interaction contrast may be obtained by multiplying the corresponding coefficients for the main effects of each factor. The interaction contrasts for all three factors compares those totals where none or exactly two factors are at high levels with those where all or exactly one factor is at a high level. One of several ways to remember the proper coefficients for a second-order interaction is to

Table 6.1   Contrast Coefficients

Coefficients for	Totals							
	$Y_{111.}$	$Y_{112.}$	$Y_{121.}$	$Y_{122.}$	$Y_{211.}$	$Y_{212.}$	$Y_{221.}$	$Y_{222.}$
A	1	1	1	1	-1	-1	-1	-1
B	1	1	-1	-1	1	1	-1	-1
C	1	-1	1	-1	1	-1	1	-1
I_{AB}	1	1	-1	-1	-1	-1	1	1
I_{AC}	1	-1	1	-1	-1	1	-1	1
I_{BC}	1	-1	-1	1	1	-1	-1	1
I_{ABC}	1	-1	-1	1	-1	1	1	-1

apply the rule: multiply the corresponding coefficients for $Y_{ijk.}$ in the three main-effects contrasts.

The expected mean squares in the AOV table apply to the model

$$Y_{ijkl} = \mu + \alpha_i + \beta_j + \gamma_k + (\alpha\beta)_{ij} + (\alpha\gamma)_{ik} + (\beta\gamma)_{jk} + (\alpha\beta\gamma)_{ijk} + \epsilon_{ijkl}$$

with

$$\alpha_1 = -\alpha_2, \quad \beta_1 = -\beta_2, \quad \gamma_1 = -\gamma_2$$

$$(\alpha\beta)_{11} = (\alpha\beta)_{22} = -(\alpha\beta)_{21} = -(\alpha\beta)_{12}$$

$$(\alpha\gamma)_{11} = (\alpha\gamma)_{22} = -(\alpha\gamma)_{21} = -(\alpha\gamma)_{12}$$

$$(\beta\gamma)_{11} = (\beta\gamma)_{22} = -(\beta\gamma)_{21} = -(\beta\gamma)_{12}$$

and

$$(\alpha\beta\gamma)_{111} = (\alpha\beta\gamma)_{122} = (\alpha\beta\gamma)_{212} = (\alpha\beta\gamma)_{221} = -(\alpha\beta\gamma)_{112} = -(\alpha\beta\gamma)_{121}$$
$$= -(\alpha\beta\gamma)_{211} = -(\alpha\beta\gamma)_{222}.$$

The expected mean squares can be used to suggest which mean squares should be compared in order to test a specified null hypotheses. Under the usual normal, independent, and homogeneous variance assumptions on ϵ_{ijkl}, the ratio of a contrast mean square to the error mean square is a Snedecor F variable. For example, suppose that it is desired to test $H_0: (\beta\gamma)_{jk} = 0$ for $j = 1, 2; k = 1, 2$ against any alternative to this hypothesis. When H_0 is true,

$$\frac{I^2_{BC}/8r}{\text{MSE}} \sim F(1, 8r - 8).$$

Hence our test rule would be: reject H_0 when

$$\frac{I^2_{BC}/8r}{\text{MSE}} > \text{tab. } F_\alpha(1, 8r - 8).$$

6.7 EXERCISES

(1) Verify the fact that the seven contrasts, created by using the coefficients displayed in Table 6.1, are mutually orthogonal.

(2) Verify the following statement made with reference to 2^3 factorial data. The coefficients obtained, by multiplying corresponding coefficients for main effects of two factors, are coefficients for a contrast in which are high-high and low-low versus high-low and low-high treatment combinations.

(3) Write out an AOV, including the expected mean square column, for the following 2^3 factorial data associated with levels of factors $\mathcal{C}$, $\mathcal{B}$, and $\mathcal{C}$.

Treatment:	(1, 1, 1)	(1, 1, 2)	(1, 2, 1)	(1, 2, 2)	(2, 1, 1)	(2, 1, 2)	(2, 2, 1)	(2, 2, 2)
	1.9	2.5	2.9	2.7	1.9	2.9	3.3	3.2
Observations:	1.9	2.7	3.2	2.5	1.8	3.1	3.1	3.1
	2.0	2.9	3.0	2.6	2.3	2.8	3.3	3.2

(4) Write out a fixed-effects model to describe the data in Exercise (3) and compute unbiased estimates of the parameters in the model.

(5) Compare the mean squares for seven orthogonal contrasts with the mean square for experimental error by computing the ratio of these quantities. State conditions under which each ratio is an observed value of a Snedecor F variable. State the null hypothesis which is being tested when the observed ratio is compared with the tabulated F value for a given Type I error.

(6) Applying the parameter restrictions suggested in the text, show that

$$\mathcal{E}\left[\frac{1}{8r} A^2\right] = \sigma^2 + 4r(\alpha^2_1 + \alpha^2_2)$$

and

$$\mathcal{E}\left[\frac{1}{8r} (I^2{}_{AB})\right] = \sigma^2 + 2r \sum_i \sum_j (\alpha\beta)^2{}_{ij}.$$

6.8 3² FACTORIAL EXPERIMENTS

The details of analyzing a 3^2 experiment are considerably different from those involved in analyzing a 2^3 experiment. A comparison study provides an interesting exercise.

Suppose that the nine treatment combinations of a 3^2 factorial experiment are replicated r times with observations denoted by Y_{ijk} and treatment combination totals denoted by $Y_{ij.}$. A first fact to consider is that now there are two orthogonal contrasts for the levels of a factor. Let the factors be denoted by $\mathcal{A}$ and $\mathcal{B}$ and direct attention to the totals for the three levels of $\mathcal{A}$. $Y_{1..} + Y_{2..} - 2Y_{3..}$ and $Y_{1..} - Y_{2..}$ are orthogonal contrasts of the totals for the levels of $\mathcal{A}$ but so are the contrasts $Y_{1..} + 2Y_{2..} - 3Y_{3..}$ and $-5Y_{1..} + 4Y_{2..} + Y_{3..}$. Still another pair of orthogonal contrasts of the totals for the levels of $\mathcal{A}$ is the pair $Y_{1..} - Y_{3..}$ and $Y_{1..} - 2Y_{2..} + Y_{3..}$. When the levels of a factor are equally spaced (the intermediate level being the average of the high and low levels), the last pair of orthogonal contrasts has a special significance.

The pooled sum of squares associated with $(t - 1)$ orthogonal contrasts of t totals has the important property that it is invariant with respect to the set of orthogonal contrasts chosen. In the present context, pairs of orthogonal contrasts $Q_1 = \displaystyle\sum_{i=1}^{t} k_i T_i$ and $Q_2 = \displaystyle\sum_{i=1}^{t} l_i T_i$ have associated with them the sum of squares

$$\frac{Q^2_1}{3\, r \displaystyle\sum_{i=1}^{3} k^2_i} \qquad \text{and} \qquad \frac{Q^2_2}{3\, r \displaystyle\sum_{i=1}^{3} l^2_i}$$

and no matter which orthogonal pair Q_1 and Q_2 is chosen, when pooled, we have

$$\frac{Q^2_1}{3\, r \displaystyle\sum_{i=1}^{3} k^2_i} + \frac{Q^2_2}{3\, r \displaystyle\sum_{i=1}^{3} l^2_i} = \sum_{i=1}^{3} \frac{(T_i - \bar{T})^2}{r}.$$

This property provides us with a principal reason for studying orthogonal contrasts instead of arbitrary sets of contrasts. Sets of orthogonal contrasts can also be identified with the data in physical and geometrical ways. The special significance of the contrasts $Y_{1..} - Y_{3..}$ and $Y_{1..} - 2Y_{2..} + Y_{3..}$ will now be studied.

If the response totals are unaffected by the level of factor $\mathcal{A}$ over the range of the levels studied, then $Y_{1..} - Y_{3..}$ will tend to be near zero. If, however, there exists a linear relationship between response and level of factor and if the slope of this relationship is appreciably different from zero, then $Y_{1..} - Y_{3..}$ will tend to depart from zero. This measure of departure from a slope of zero is often called the linear A contrast although it is not, strictly speaking, a measure of linearity. It will be denoted by A_l. If the response is a quadratic function of the level, the contrast $Y_{1..} - 2Y_{2..} + Y_{3..}$ tends to be different from zero and acts as a measure of the departure from linearity. It is called the quadratic A contrast and A_q will be used to denote it.

Similar orthogonal contrasts can be computed from the totals for the three levels of factor ℬ. These contrasts and those for α are mutually orthogonal. Furthermore, these contrasts can be augmented with one of infinitely many sets of four contrasts to form a set of eight mutually orthogonal contrasts of the nine treatment combination totals $Y_{ij.}$. The augmented contrasts will not be contrasts of levels of α or ℬ alone. They correspond to the four degrees of freedom attached to the interaction of factors A and ℬ. Since the coefficients for four orthogonal interaction contrasts can readily be obtained by multiplying coefficients for α contrasts by those for ℬ contrasts, this is the method usually used to obtain interaction coefficients. The coefficients obtained by multiplying those for linear A by those for linear B furnish the set for an interaction which will be denoted by I_{ll}. I_{lq}, I_{ql}, and I_{qq} will denote the other correspondingly constructed contrasts. A table of these coefficients follows. The numbers in Table 6.2 are the coefficients that should be attached to the totals $Y_{ij.}$. Notice that contrasts for levels of α(ℬ) alone could be expressed alternatively in terms of the totals $Y_{i..}(Y_{.j.})$.

Table 6.2   Contrast Coefficients for a 3² Factorial Experiment

Treatment combination totals

Coefficients k_{ij} for	$Y_{11.}$	$Y_{12.}$	$Y_{13.}$	$Y_{21.}$	$Y_{22.}$	$Y_{23.}$	$Y_{31.}$	$Y_{32.}$	$Y_{33.}$	$\lambda = \sum_{j} \sum_{i} k^2_{ij}$
A_l	1	1	1	0	0	0	-1	-1	-1	6
A_q	1	1	1	-2	-2	-2	1	1	1	18
B_l	1	0	-1	1	0	-1	1	0	-1	6
B_q	1	-2	1	1	-2	1	1	-2	1	18
I_{ll}	1	0	-1	0	0	0	-1	0	1	4
I_{lq}	1	-2	1	0	0	0	-1	2	-1	12
I_{ql}	1	0	-1	-2	0	2	1	0	-1	12
I_{qq}	1	-2	1	-2	4	-2	1	-2	1	36

The coefficient table affords a systematic way of calculating not only the value of a contrast but also the sum of squares associated with individual degrees of freedom. A detailed breakdown of the treatment sum of squares in an r replicated 3² experiment follows. It is one means by which some aspects of the αℬ interaction can be brought to light. Soon another look at interaction will be taken from the model point of view. The AOV approach involves the use of a parameter λ. The definition and some needed values of this parameter are found in the last column of the coefficient table.

Source	df	SS
Treatment combinations	8	$\dfrac{1}{r}\displaystyle\sum_{j}\sum_{i} Y^2{}_{ij.} - \dfrac{Y^2{}_{...}}{9r}$
Level of $\mathcal{C}$	2	$\dfrac{1}{3r}\displaystyle\sum_{i} Y^2{}_{i..} - \dfrac{Y^2{}_{...}}{9r}$
$\mathcal{C}_l$	1	$\dfrac{1}{r\lambda_l} A^2{}_l$
$\mathcal{C}_q$	1	$\dfrac{1}{r\lambda_q} A^2{}_q$
Level of $\mathcal{B}$	2	$\dfrac{1}{3r}\displaystyle\sum_{j} Y^2{}_{.j.} - \dfrac{Y^2{}_{...}}{9r}$
$\mathcal{B}_l$	1	$\dfrac{1}{r\lambda_l} B^2{}_l$
$\mathcal{B}_q$	1	$\dfrac{1}{r\lambda_q} B^2{}_q$
Interaction	4	$\dfrac{1}{r}\displaystyle\sum_{j}\sum_{i} Y^2{}_{ij.} - \dfrac{1}{3r}\displaystyle\sum_{j} Y^2{}_{.j.}$ $- \dfrac{1}{3r}\displaystyle\sum_{i} Y^2{}_{.i.} + \dfrac{Y^2{}_{...}}{9r}$
Linear by linear	1	$\dfrac{1}{r\lambda_{ll}} I^2{}_{ll}$
$\mathcal{C}$ linear by $\mathcal{B}$ quadratic	1	$\dfrac{1}{r\lambda_{lq}} I^2{}_{lq}$
$\mathcal{C}$ quadratic by $\mathcal{B}$ linear	1	$\dfrac{1}{r\lambda_{ql}} I^2{}_{ql}$
Quadratic by quadratic	1	$\dfrac{1}{r\lambda_{qq}} I^2{}_{qq}$
Experimental error	$(9r - 1)$	pooled within treatment sum of squares

Consider now a model to explain the observation Y_{ijk}. With $\epsilon_{ijk} \sim$ NID$(0, \sigma^2)$ and all other parameters fixed, Y_{ijk} is often hypothesized to be $Y_{ijk} = \mu + \alpha_i + \beta_j + \gamma_{ij} + \epsilon_{ijk}$, $i, j = 1, 2, 3$. This is a satisfactory model for some purposes with perhaps its main drawback being that it greatly overparameterizes the situation. One set of restrictions that yields a workable model is the following. Let μ, α_1, α_3, β_1, β_3, γ_{11}, γ_{13}, γ_{31}, and γ_{33} be mathematically independent and let $\alpha_2 = -(\alpha_1 + \alpha_3)$, $\beta_2 = -(\beta_1 +$

β_3), $\gamma_{i2} = -(\gamma_{i1} + \gamma_{i3})$, and $\gamma_{2j} = -(\gamma_{1_j} + \gamma_{3j})$. These restrictions leave all parameters estimable and yet allow for an adequate description of data by the model $Y_{ijk} = \mu + \alpha_i + \beta_j + \gamma_{ij} + \epsilon_{ijk}$.

Before discussion of this model for 3^2 factorial data proceeds further, an illustration will be considered in some detail. Suppose that factor α has a quadratic response relationship with respect to levels of the factor. For illustrative purposes, let $\alpha_1 = -2$ and $\alpha_3 = 6$. Suppose that factor β contributed to the response in a linear fashion; that is, $\beta_1 = -1$ and $\beta_3 = 1$. Suppose also that $\mu = 20$ with $\gamma_{11} = 1$, $\gamma_{13} = 2$, $\gamma_{31} = 4$, and $\gamma_{33} = -6$. Applying the parameter restrictions described earlier, the values of all the parameter, except that for σ^2, are exhibited in the arrays

$$\begin{bmatrix} \gamma_{11} & \gamma_{12} & \gamma_{13} \\ \gamma_{21} & \gamma_{22} & \gamma_{23} \\ \gamma_{31} & \gamma_{32} & \gamma_{33} \end{bmatrix} = \begin{bmatrix} 1 & -3 & 2 \\ -5 & 1 & 4 \\ 4 & 2 & -6 \end{bmatrix}, \begin{bmatrix} \alpha_1 \\ \alpha_2 \\ \alpha_3 \end{bmatrix} = \begin{bmatrix} -2 \\ -4 \\ 6 \end{bmatrix}, \quad \text{and}$$

$$\begin{bmatrix} \beta_1 \\ \beta_2 \\ \beta_3 \end{bmatrix} = \begin{bmatrix} -1 \\ 0 \\ 1 \end{bmatrix}.$$

In this illustration, r was set equal to 2 and the sets of observations were created from the model by rounding off to one decimal place the values of ϵ_{ijk} read from the tables of normal random errors. The errors ϵ_{ijk} were selected with variance 0, 2, and 32 with the specific values of ϵ_{ijk} for $\sigma^2 = 32$ obtained from those with $\sigma^2 = 2$ through multiplying by 4. The created data for the three situations appear in Table 6.3.

	Values of $\mu + \alpha_i + \beta_j$					*Values of Y_{ijk} when $\sigma^2 = 0$ or, equivalently, $\mu + \alpha_i + \beta_j + \gamma_{ij}$*		
α\β	*Low*	*Medium*	*High*		α\β	*Low*	*Medium*	*High*
Low	17 17	18 18	19 19		*Low*	18 18	15 15	21 21
Medium	15 15	16 16	17 17		*Medium*	10 10	17 17	21 21
High	25 25	26 26	27 27		*High*	29 29	28 28	21 21

A careful study of the analysis of variance for these sets of data can be very revealing. A few of the more striking features are mentioned here. First, β was chosen to be linear but, after applying the errors with variance equal to 32, the β linear contrast yielded zero leaving all of the β sum of squares in the quadratic contrast. When no error was

Table 6.3   Created Data for Two Error Variances

Values of Y_{ijk} when $\sigma^2 = 2$ Values of Y_{ijk} when $\sigma^2 = 32$

$\alpha \backslash \mathcal{B}$	Low	Medium	High	$\alpha \backslash \mathcal{B}$	Low	Medium	High
Low	21.1 19.2	15.2 16.0	22.6 18.6	Low	30.4 22.8	15.8 19.0	27.4 11.4
Medium	10.7 12.3	18.1 14.3	21.2 21.5	Medium	12.8 19.2	21.4 6.2	21.8 23.0
High	27.2 28.2	26.8 25.8	20.9 22.9	High	21.8 25.8	23.2 19.2	20.6 28.6

applied, each nonzero mean square was of course significantly different from zero. Most mean squares remained significant when errors with variance equal to 2 were applied, but when σ^2 was chosen to be 32, the errors almost completely disguise the effects and interaction, causing none of the mean squares to be declared significantly different from zero, and thus pointing up the need for error control in experimental design work.

AOV

Source	df	When $\sigma^2 = 0$ SS	MS	When $\sigma^2 = 2$ SS	MS	When $\sigma^2 = 32$ SS	MS
Total	18	7772.00		7747.36		8258.56	
Mean	1	7200.00		7304.38		7622.01	
Treatments	8	572.00	71.50**	421.31	52.66**	289.83	36.23
α	2	336.00	168.00**	256.98	128.49**	103.70	51.85
α_l	1	192.00	192.00**	127.40	127.40**	12.81	12.81
α_q	1	144.00	144.00**	129.58	129.58**	90.88	90.88
$\mathcal{B}$	2	12.00	6.00**	12.19	6.10	87.11	43.55
$\mathcal{B}_l$	1	12.00	12.00**	6.75	6.75	.00	.00
$\mathcal{B}_q$	1	.00	.00	5.44	5.44	87.11	87.11
Interaction	4	224.00	56.00**	152.14	38.03**	99.02	24.75
I_{ll}	1	60.50	60.50**	19.53	19.53*	32.00	32.00
I_{lq}	1	37.50	37.50**	26.25	26.25**	4.51	4.51
I_{ql}	1	121.50	121.50**	104.58	104.58**	61.44	61.44
I_{qq}	1	4.50	4.50**	1.77	1.77	1.08	1.08
Experimental error	9	.00	.00	21.77	2.42	346.72	38.52

* Indicates significance at the 5% level.
** Indicates significance at the 1% level.

The errors chosen in order to create the data, when σ^2 was selected to be 2, appear in the array:

$\alpha\backslash\beta$	Low	Medium	High
Low	3.1 1.2	.2 1.0	1.6 -2.4
Medium	.7 2.3	1.1 -2.7	.2 .5
High	-1.8 $-.8$	-1.2 -2.2	$-.1$ 1.9

The average value of these errors is $\bar{\epsilon} = .2$ and the sample variance s^2_ϵ based on 17 degrees of freedom is 2.826. Remember that, in actual analysis of compiled data, the errors are not available and estimation of σ^2 must be accomplished with fewer degrees of freedom. In our illustration, $\hat{\sigma}^2$ is based on nine degrees of freedom.

Let us now see how well, from the created data, we can estimate the parameters with which we started. Adding all observations, we obtain

$$Y_{\ldots} = 9r\mu + 3r \sum_{i=1}^{3} \alpha_i + 3r \sum_{j=1}^{3} \beta_j + r \sum_i \sum_j \gamma_{ij} + \epsilon \cdots$$

Table 6.4

Parameter	Hypothesized value	Estimate when $\sigma^2 = 2$	Estimate when $\sigma^2 = 32$
μ	20	20.14	20.58
α_1	-2	-1.36	.56
α_2	-4	-3.80	-3.18
α_3	6	5.16	2.62
β_1	-1	$-.36$	1.55
β_2	0	$-.78$	-3.10
β_3	1	1.14	1.55
γ_{11}	1	1.73	3.91
γ_{12}	-3	-2.41	$-.62$
γ_{13}	2	.68	-3.29
γ_{21}	-5	-4.49	-2.96
γ_{22}	1	.63	$-.49$
γ_{23}	4	3.86	3.44
γ_{31}	4	2.76	$-.16$
γ_{32}	2	1.78	1.11
γ_{33}	-6	-4.54	$-.96$
σ^2	2 and 32	2.42	38.52

Since $\Sigma\alpha_i = \Sigma\beta_j = \Sigma\Sigma\gamma_{ij} = 0$, $Y_{...}/9r = \mu + \epsilon_{...}/9r$. The expected value of $\epsilon_{...}/9r$ is zero and its variance is $\sigma^2/9r$. Therefore, $Y_{...}/9r$ is an unbiased estimate of μ with variance $\sigma^2/9r$. In similar manners $\hat{\alpha}_i = Y_{i..}/3r - Y_{...}/9r$, $\hat{\beta}_j = Y_{.j.}/3r - Y_{...}/9r$, and $\hat{\gamma}_{ij} = Y_{ij.}/r - Y_{i..}/3r - Y_{.j.}/3r + Y_{...}/9r$ can be shown to be unbiased estimates of α_i, β_j, and γ_{ij}. σ^2 is estimated in the usual way from the experimental error mean square. Table 6.4 gives the estimated values of the parameters along with the hypothesized values.

Table 6.4 accentuates the estimation encumbrance that errors with large variance bring forth. This table concludes the discussion of the illustrative example.

6.9   THE ROLE OF FACTORIAL EXPERIMENTATION

What has been done in the previous discussion, indeed for the most part throughout the text, was to analyze models. This seems proper, for experimentation seems to make headway when adequate models from the description point of view are developed followed by a sufficient development of techniques for solving model problems.

Consider an experimenter who wishes to devise an experiment because of some thoughts, some accidents, and/or some past experiments. Before the experiment is run, some thought should be given to how the data is to be obtained and how it is to be analyzed. There was a time when advice from experts dictated the holding constant of the levels of all factors except one, until enough was learned about the factor that was allowed to vary. The procedure then moved on to a second factor while the first factor, along with the rest, was held constant. Today this advice is seldom given because it is realized that interactions of factors are, in most experiments, very important. The one-factor-at-a-time method is now understood to be wasteful with respect to time spent experimenting and experimental material used. The one-factor-at-a-time method may even lead to some misleading conclusions when employed in search-type experimentation.

Suppose now that the experimenter has been exposed to statistics in degree comparable to that of a student of this text. His exposure to 2^3 and 3^2 experimentation might lead him to devise his experiment in such a way that his treatment combinations have one of these structures; having done this, he might then be led to adopt a model discussed in this chapter. These actions might prompt several questions for a thinking bystander. Should a factorial arrangement to treatments be used? Does the model discussed here adequately describe the data? Is it not a poor policy to use methods simply because they are known by the experimenter? Is a little knowledge of experimental design a dangerous thing?

Each question has merit and perhaps no final answer. This is part of the beauty of statistics. Experiments are like people, in that no two are alike. But generally the experimenter in a situation such as that described here would do well to draw upon his exposure to factorial experimentation. Some appropriate remarks follow which support this view. Let it be remembered that much experimentation has been rather haphazard. Factorial experimentation is a big step in making data gathering systematic. Analysis of data, obtained from experimental units that were assigned treatments in nonfactorial fashions, may become far more complicated than any methods considered thus far. The application of a better-fitting model may also lead to an impossible or extremely detailed analysis. Of course, we do want the model to adequately fit the data and there are tests for fit, but experience with data is often required to properly interpret the test. The model discussed here and closely related models discussed in a later chapter have proved to be adequate for many interaction-type experiments and are in wide use today. Thus the 2^3 and 3^2 factorial models become important because they can be included in an experimenter's repertoire of easily analyzed models. They are important, too, because they exhibit basic principles that underlie many of the experimental design models. Most of the computing details for balanced l^j factorial experiments follow principles encountered in the analysis of 2^3 and 3^2 factorial data.

The treatment combinations in each discussion and example in this chapter were assigned to experimental units at random. Thus we always were studying factorial arrangement of treatments in a completely randomized design. When the term "block" did appear, it was used in the sense of a factor with the consequence that treatment i in block j was to be thought of as a treatment combination. Factorial experiments can also be run in statistical designs other than the completely randomized design, with the important idea to keep in mind being that "treatment combination" now plays the role that treatment used to play. The situation when unequal numbers of replicates occur for different treatment combinations is conceptually and computationally more complicated.

The development of the notion of studying multiple characteristics of experimental units can take at least two directions. In Chapter 5 one characteristic was studied with the hope that it would shed some light on the response characteristic. We called this type of study a regression analysis. In Chapter 6 we studied the responses to various prechosen symmetric sets of characteristics (levels of factors) in order to shed some light on the relationships among factors. Fundamentally, the theory in

Chapter 6 was a generalization of the theory in Chapter 5. If thought about in the proper way, every experimental design model can be phased as a regression-type model. To the experimental design statistician, regression analysis is employed so that our knowledge about one factor might be used to reduce the experimental error variance, thereby strengthening statements made about a second factor.

6.10 EXERCISES

(1) Let $T_1 T_2$ and T_3 be the response totals associated with the three levels of a factor. Let $Q_1 = \sum_{i=1}^{3} k_i T_i = T_1 - T_3$, $\lambda_1 = \sum_{i=1}^{3} k^2_i = 2$, $Q_2 = \sum_{i=1}^{3} l_i T_i = T_1 - 2T_2 + T_3$, $\lambda_2 = \sum_{i=1}^{3} k^2_i = 6$, $Q^*_1 = \sum_{i=1}^{3} k^*_i T_i = T_1 + 2T_2 - 3T_3$, $\lambda^*_1 = \sum_{i=1}^{3} k^*_i = 14$, $Q^*_2 = \sum_{i=1}^{3} l^*_i T_i = -5T_1 + 4T_2 + T_3$, and $\lambda^*_2 = \sum_{i=1}^{3} l^*_i = 42$. Show that

$$\frac{Q^2_1}{\lambda_1} + \frac{Q^2_2}{\lambda_2} = \frac{Q^{*^2}_1}{\lambda^*_1} + \frac{Q^{*^2}_2}{\lambda^*_2}$$

(2) Because the quadratic contrast A_q involves only levels of a factor α, it can be written in the following two ways:

$$A_q = \sum_{j=1}^{3} \sum_{i=1}^{3} k_{ij} Y_{ij.} = Y_{11.} + Y_{12.} + Y_{13.} - 2Y_{21.} - 2Y_{22.} - 2Y_{23.}$$
$$+ Y_{31.} + Y_{32.} + Y_{33.}$$

$$A_q = \sum_{i=1}^{3} l_i Y_{i..} = Y_{1..} - 2Y_{2..} + Y_{3...}$$

Show that the corresponding two expressions for obtaining the sum of squares associated with A quadratic are equivalent. That is, show that

$$\frac{\left(\sum_j \sum_i k_{ij} Y_{ij.}\right)^2}{r \left(\sum_j \sum_i k^2_{ij}\right)} = \frac{\left(\sum_i l_i Y_{i..}\right)^2}{3r \left(\sum_i l_i\right)^2}$$

(3) In order to obtain a feeling for the magnitude of contrasts and their corresponding sum of squares, compute these quantities for the linear and quadratic contrasts in the following response situations (Figure 6.2).

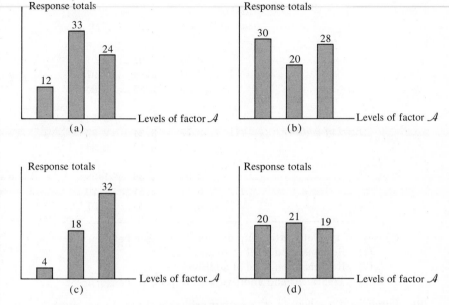

Figure 6.2

(4)   Verify all calculation for the 3^2 factorial illustrative example of Section 6.8.

(5)   Show that $\hat{\alpha}_i = Y_{i..}/3r - Y.../9r$ is an unbiased estimator of α_i and determine the variance of $\hat{\alpha}_i$.

(6)   Verify that the contrasts denoted by A_l, A_q, B_l, B_q, I_{ll}, I_{lq}, I_{ql}, and I_{qq} are mutually orthogonal.

(7)   Examine the following facts:
   (a)   If α_1 and α_3 are chosen to be -2 and 6 and $\alpha_1 + \alpha_2 + \alpha_3 = 0$ is required then the factor $\mathcal{C}$ has a nonlinear response relationship with respect to levels of $\mathcal{C}$.
   (b)   If the contribution to the response made by factor $\mathcal{B}$ is a linear function of the level of $\mathcal{B}$ and if $\beta_1 + \beta_2 + \beta_3$ is required to be zero then β_2 must be zero.
   Construct graphs of these and other situations involving three parameters.

(8)   Consider three replicates of 3^2 factorial data.

   (a)   Write an appropriate model for the data and state assumptions that are sufficient for testing (1) equality of the effects of different levels of the treatment, and (2) the hypothesis of negligible interaction of treatment and additive.
   (b)   Write out an AOV and use it as an aid to test the above hypotheses.

	Level of treatment	I	II	III
		2.14	2.91	3.01
	I	2.09	2.74	2.97
		2.30	2.84	3.04
		2.90	3.07	2.54
Level of additive	II	2.77	3.24	2.37
		2.85	3.12	2.34
		2.95	2.38	1.98
	III	3.11	2.51	2.20
		3.19	2.43	2.07

(c) What type of relationship do you think exists between:
   (1) Response and level of treatment?
   (2) Response and level of additive?
   (3) Treatment and additive with respect to the response?

(9) Throughout a school term, a certain high-school athlete was given a certain strength test each Friday. For 48 hours before each test, all proteins and sugars consumed by the boy were carefully measured and controlled. Nine equally spaced treatment combinations of protein and sugar levels were arranged factorially and treatment combinations were assigned to weeks of the academic term in a completely randomized fashion in order to hopefully minimize the effects of any trend in strength response. Assuming that Y_{ijk}, the k response to the ith level of protein and the jth level of sugar can be adequately described by the model

$$Y_{ijk} = \mu + \alpha_i + \beta_j + \gamma_{ij} + \epsilon_{ijk}$$

with the usual parameter restrictions, estimate the following:
(a) α_i; $i = 1, 2, 3$, the main effect of the ith level of protein.
(b) β_j; $j = 1, 2, 3$, the main effect of the jth level of sugar.

	Protein	Low	Medium	High
	Low	58	51	63
		71	67	65
Sugar	Medium	68	72	69
		71	79	80
	High	80	84	83
		63	65	81

(c) γ_{33}; the interaction effect of the high levels of both protein and sugar.

(d) σ^2; the variance of $\epsilon_{ijk} \sim \mathrm{NID}(0, \sigma^2)$.

(e) Are any of the main effects significantly different from zero?

(10) A factorial experiment where there are two levels of one factor and three levels of a second factor is called a 2×3 experiment. Fill out the AOV for the following data associated with a 2×3 factorial arrangement of treatments run in a completely randomized setting.

	Levels of factor ⑥	Low	Medium	High
		31	29	34
	Low	34	27	31
		36	30	34
Levels of factor ⑧		33	28	30
	High	35	27	27
		36	28	29

AOV

Source of variation	df	SS	MS
Total	18	—	
Mean	1	—	
Treatment combinations	—	—	—
Levels of ⑧	—	—	—
Levels of ⑥	—	—	—
⑥ linear	—	—	—
⑥ quadratic	—	—	—
Interaction of ⑧ and ⑥	—	—	—
⑧ by ⑥ linear	—	—	—
⑧ by ⑥ quadratic	—	—	—
Experimental error	—	—	—

7

STATISTICAL
INFERENCE FOR SOME
SPECIAL MODELS

7.1 INTRODUCTION

A prominent statistician once described his experience with analysis of variance. He said, "When I took my first course in statistics I thought I would never learn all there was to know about analysis of variance." He continued, "During my second course I decided that I knew all there was to know about the topic but, during my last course, I decided again I would never learn all there was to know about analysis of variance." One reason for the inclusion of this chapter is that we wish to throw open the doors to new areas of analysis of variance with the hope that the student, after finishing the book, will not feel as if he had exhausted the subject. There are, of course, many other good reasons for including a study of some special models. Several of the models discussed are in common use in many experimental areas. The study of hierarchal random-effects models seems to complement nicely the study of cross-classification fixed-effects models.

The first concept that will be expanded was introduced in Section 4.4. It was presented in association with a model of the type

$$Y_{ij} = \mu + \alpha_i + \beta_{ij}$$

where $\alpha_i \sim N(0, \sigma^2_\alpha)$, $\beta_{ij} \sim N(0, \sigma^2_\beta)$, and μ is a fixed unknown parameter. We consider here the balanced case; that is, the case where $i = 1$,

$2, \ldots, a$ and $j = 1, 2, \ldots, b$. We assume, too, that all α_i and β_{ij} are independent random variables.

We now call attention to a fundamental difference between this model, and the fixed-effects models. In the fixed-effects models, the random component of the observation was always assumed to be independent of the random component for any other observation. This is certainly not the case for all observations in the random-effects model. For $j \neq j'$,

$$\text{Cov } (Y_{ij}, Y_{ij'}) = \mathcal{E}[(Y_{ij} - \mu)(Y_{ij'} - \mu)] = \mathcal{E}[(\alpha_i + \beta_{ij})(\alpha_i + \beta_{ij'})]$$
$$\text{Cov } (Y_{ij}, Y_{ij'}) = \mathcal{E}[\alpha^2{}_i] + \mathcal{E}[\alpha_i\beta_{ij}] + \mathcal{E}[\alpha_i\beta_{ij'}] + \mathcal{E}[\beta_{ij}\beta_{ij'}].$$

Applying the independence assumptions, we have

$$\text{Cov } (Y_{ij}, Y_{ij'}) = \sigma^2{}_\alpha.$$

Now for $i \neq i'$,

$$\text{Cov } (Y_{ij}, Y_{i'j}) = \mathcal{E}[(\alpha_i + \beta_{ij})(\alpha_{i'} + \beta_{i'j})].$$
$$\text{Cov } (Y_{ij}, Y_{i'j}) = \mathcal{E}[\alpha_i\alpha_{i'}] + \mathcal{E}[\alpha_i\beta_{i'j}] + \mathcal{E}[\alpha_{i'}\beta_{ij}] + \mathcal{E}[\beta_{ij}\beta_{i'j}].$$
$$\text{Cov } (Y_{ij}, Y_{i'j}) = 0.$$

Let us think of the observations Y_{ij} as belonging to one of "a" categories of a classification scheme A, and let B_{ij} represent the j experimental unit in category i of classification A. Y_{ij} is the observation made on experimental unit B_{ij}.

$$
\begin{array}{ccccc}
\underline{A_1} & \cdots & \underline{A_i} & \cdots & \underline{A_a} \\
B_{11} & & B_{i1} & & B_{a1} \\
\cdot & & \cdot & & \cdot \\
\cdot & & \cdot & & \cdot \\
\cdot & & \cdot & & \cdot \\
B_{1j} & \cdots & B_{ij} & \cdots & B_{aj} \\
\cdot & & \cdot & & \cdot \\
\cdot & & \cdot & & \cdot \\
\cdot & & \cdot & & \cdot \\
B_{1b} & \cdots & B_{ib} & \cdots & B_{ab}
\end{array}
$$

This scheme of classification will be called a one-way or *one-fold hierarchy* and it is a very special case of an n-way or *n-fold hierarchy*. Any two observations in the same category of classification A have covariance equal to $\sigma^2{}_\alpha$.

The following AOV for a one-fold hierarchy can be used to illustrate the basic partitioning concept for all hierarchal situations.

Source of variation	df	SS	$\mathcal{E}MS$
Among B	$ab - 1$	$\sum_{ij} Y^2_{ij} - Y^2../ab$	
Within A	$a(b - 1)$	$\sum_{ij} Y^2_{ij} - \sum_i Y^2_{i.}/b$	σ^2_β
Among A	$a - 1$	$\sum_i Y^2_{i.}/b - Y^2../ab$	$\sigma^2_\beta + b\sigma^2_\alpha$

Among B is a shortened form for "among all subclassification categories."

Within A is a shortened form for "pooled within categories of main classification."

Among A is a shortened form for "among all categories of main classification."

The relationship *Among subclassification = Within plus Among* will be called the basic partitioning rule.

The expected mean square column is again a very important part of the AOV. Consider the within A source of variation.

$$\sum_{ij} Y^2_{ij} - \frac{\sum_i Y^2_{i.}}{b} = \sum_i \left[\sum_j Y^2_{ij} - \frac{Y^2_{i.}}{b} \right] = \sum_i \left[\sum_j \left(Y_{ij} - \frac{Y_{i.}}{b} \right)^2 \right]$$

Now $Y_{i.}/b = \mu + \alpha_i + \sum_j \beta_{ij} / b$ so that

$$\sum_j \left(Y_{ij} - \frac{Y_{i.}}{b} \right)^2 = \sum_j (\beta_{ij} - \bar\beta_i)^2. \quad \mathcal{E}\left[\frac{1}{b-1} \sum_j (\beta_{ij} - \bar\beta_i)^2 \right] = \sigma^2_\beta$$

hence

$$\mathcal{E}\left\{ \frac{1}{a(b-1)} \left[\sum_{ij} Y^2_{ij} - \frac{\sum_i Y^2_{i.}}{b} \right] \right\} = \frac{1}{a} \sum_i \mathcal{E}\left[\frac{1}{b-1} \sum_j (\beta_{ij} - \bar\beta_i)^2 \right]$$

$$= \frac{1}{a} \sum_i \sigma^2_\beta = \sigma^2_\beta.$$

Another way of viewing the situation involves showing that

$$\frac{\sum_i \left[\sum_j (Y_{ij} - Y_{i.}/b)^2 \right]}{\sigma^2_\beta} \sim \chi^2[a(b-1)].$$

Since the expected value of a Chi Square variable is its degrees of freedom, we have $\mathcal{E}\left\{\left[\sum\limits_{ij} Y^2_{ij} - \sum\limits_{i} Y^2_{i.}/b\right]\Big/\sigma^2_\beta\right\} = a(b-1)$ from which it follows immediately that the $\mathcal{E}MS$ for Within $A = \sigma^2_\beta$. The details, related to showing that the $\mathcal{E}MS$ for Among $A = \sigma^2_\beta + b\sigma^2_\alpha$, are left as exercises.

As was mentioned in Chapter 4, unbiased estimates of σ^2_β and σ^2_α usually are obtained by first equating mean squares to expected mean squares and then solving the resulting equations.

7.2   THE TWO-FOLD HIERARCHAL MODEL

Consider now the balanced two-fold hierarchal situation.

A_1			A_i			A_a		
B_{11} $\cdots$	B_{1j} $\cdots$	B_{1b}	B_{i1} $\cdots$	B_{ij} $\cdots$	B_{ib}	B_{a1} $\cdots$	B_{aj} $\cdots$	B_{ab}
C_{111} $\cdots$	C_{1j1} $\cdots$	C_{1b1}	C_{i11} $\cdots$	C_{ij1} $\cdots$	C_{ib1}	C_{a11} $\cdots$	C_{aj1} $\ldots$	C_{ab1}
.	.	.	.	.	.	.	.	.
.	.	.	.	.	.	.	.	.
C_{11k}	C_{1jk}	C_{1bk}	C_{i1k}	C_{ijk}	C_{ibk}	C_{a1k}	C_{ajk}	C_{abk}
.	.	.	.	.	.	.	.	.
.	.	.	.	.	.	.	.	.
C_{11c} $\cdots$	C_{1jc} $\cdots$	C_{1bc}	C_{i1c} $\cdots$	C_{ijc} $\cdots$	C_{ibc}	C_{a1c} $\cdots$	C_{ajc} $\cdots$	C_{abc}

In this classification scheme C_{ijk} denotes the kth experimental unit classified according to the jth category of classification B in the ith category of classification A. Denote the observation on experimental unit C_{ijk} by Y_{ijk} and consider the model

$$Y_{ijk} = \mu + \alpha_i + \beta_{ij} + \gamma_{ijk}$$

where μ is a fixed unknown parameter, $\alpha_i \sim N(0, \sigma^2_\alpha)$, $\beta_{ij} \sim N(0, \sigma^2_\beta)$, $\gamma_{ijk} \sim N(0, \sigma^2_\gamma)$ and all α_i, β_{ij}, and γ_{ijk} are independent.

α_i can be thought of as the random effect peculiar to the ith category of A.

$\alpha_i + \beta_{ij}$ can be thought of as the random effect peculiar to jth category of B in the ith category of A.

$\alpha_i + \beta_{ij} + \gamma_{ijk}$ is the random contribution associated with experimental unit C_{ijk}.

In this situation, two observations in different categories of the main classification are uncorrelated. Observations in the same category of A but different categories of B have covariance equal to σ^2_α and observations in the same category of A and the same category of B have covariance equal to $\sigma^2{}_\alpha + \sigma^2{}_\beta$.

By using the basic partitioning rule, we can easily write out a rather helpful and complete AOV.

$$\text{Among } C = C\text{'s Within } B + \text{Among } B.$$

We can next partition Among B.

$$\text{Among } B = B\text{'s Within } A + \text{Among } A.$$

When all sums of squares are put on a per-observation basis the AOV is:

Source of variation	df	SS	$\mathcal{E}MS$
Among C	$abc - 1$	$\displaystyle\sum_{ijk} Y^2{}_{ijk} - Y^2\ldots/abc$	
C's Within B	$ab(c - 1)$	$\displaystyle\sum_{ijk} Y^2{}_{ijk} - \sum_{ij} Y^2{}_{ij.}/c$	$\sigma^2\gamma$
Among B $\Big\{$ B's Within A	$ab - 1$ $\Big\{$ $a(b - 1)$	$\displaystyle\sum_{ij} Y^2{}_{ij.}/c - \sum_{i} Y^2{}_{i..}/bc$	$\sigma^2\gamma + c\sigma^2{}_\beta$
Among A	$a - 1$	$\displaystyle\sum_{i} Y^2{}_{i..}/bc - Y^2\ldots/abc$	$\sigma^2\gamma + c\sigma^2{}_\beta + bc\sigma^2{}_\alpha$

A table of the necessary observation totals is helpful in obtaining the sums of squares.

B_{11} $\cdots$ B_{1j} $\cdots$ B_{1b}	B_{i1} $\cdots$ B_{ij} $\cdots$ B_{ib}	B_{a1} $\cdots$ B_{aj} $\cdots$ B_{ab}
Y_{111} $\quad$ Y_{1j1} $\quad$ Y_{1b1}	Y_{i11} $\quad$ Y_{ij1} $\quad$ Y_{ib1}	Y_{a11} $\quad$ Y_{aj1} $\quad$ Y_{ab1}
$\cdot$ $\quad$ $\cdot$ $\quad$ $\cdot$	$\cdot$ $\quad$ $\cdot$ $\quad$ $\cdot$	$\cdot$ $\quad$ $\cdot$ $\quad$ $\cdot$
Y_{11k} $\quad$ Y_{1jk} $\quad$ Y_{1bk}	Y_{i1k} $\quad$ Y_{ijk} $\quad$ Y_{ibk}	Y_{ajk} $\quad$ Y_{ajk} $\quad$ Y_{abk}
$\cdot$ $\quad$ $\cdot$ $\quad$ $\cdot$	$\cdot$ $\quad$ $\cdot$ $\quad$ $\cdot$	$\cdot$ $\quad$ $\cdot$ $\quad$ $\cdot$
Y_{11c} $\quad$ Y_{1jc} $\quad$ Y_{1bc}	Y_{i1c} $\quad$ Y_{ijc} $\quad$ Y_{ibc}	Y_{a1c} $\quad$ Y_{ajc} $\quad$ Y_{abc}
Totals $Y_{11.}$ $\quad$ $Y_{1j.}$ $\quad$ $Y_{1b.}$	$Y_{i1.}$ $\quad$ $Y_{ij.}$ $\quad$ $Y_{ib.}$	$Y_{a1.}$ $\quad$ $Y_{aj.}$ $\quad$ $Y_{ab.}$
$Y_{1.}$	$Y_{i..}$	$Y_{a..}$

$$Y\ldots$$

The table also suggests the degrees of freedom for each source of variability, and the degrees of freedom in turn suggest the set of observational totals that go into each of the sums of squares. The coefficients of

$\sigma^2{}_\beta$ and $\sigma^2{}_\alpha$ in the $\mathcal{E}MS$ column can be obtained from the table of observations by remembering that the coefficient of $\sigma^2{}_\beta$ is the number of times a particular β_{ij} appears among the (abc) observations and the coefficient of $\sigma^2{}_\alpha$ is the number of times a particular α_i appears among the (abc) observations.

Some experimenters prefer to display the AOV in the following form:

Source of variation	df	SS	$\mathcal{E}MS$
Total corrected	$abc - 1$	$\sum_{ijk} Y^2{}_{ijk} - Y^2{}_{...}/abc$	
Among A	$a - 1$	$\sum_{i} Y^2{}_{i..}/bc - Y^2{}_{...}/abc$	$\sigma^2{}_\gamma + c\sigma^2{}_\beta + bc\sigma^2{}_\alpha$
B's in A	$a(b - 1)$	$\sum_{ij} Y^2{}_{ij.}/c - \sum_{i} Y^2{}_{i..}/bc$	$\sigma^2{}_\gamma + c\sigma^2{}_\beta$
C's in B	$ab(c - 1)$	$\sum_{ijk} Y^2{}_{ijk} - \sum_{ij} Y^2{}_{ij.}/c$	$\sigma^2{}_\gamma$

Point estimates of $\sigma^2{}_\gamma$, $\sigma^2{}_\beta$, and $\sigma^2{}_\alpha$ again usually are obtained by solving the equations resulting from setting the mean squares equal to the expected mean squares. Confidence interval estimates are not so easily obtained and will not be discussed here. Tests of hypotheses are difficult or easy, depending on the nature of the hypothesis to be tested. Fortunately, many of the hypotheses of interest are of the easy variety. In order to illustrate hypothesis testing, consider the problem of testing $H_0: \sigma^2{}_\beta = 0$ against $\sigma^2{}_\beta > 0$. The test statistic (as is suggested by the $\mathcal{E}MS$ column) is

$$\frac{MS(B\text{'s in } A)}{MS(C\text{'s in } B)}.$$

This follows because of the fact that

$$\frac{\sum_{ij} Y^2{}_{ij.}/c - \sum_{i} Y^2{}_{i..}/bc}{a(b - 1)[\sigma^2{}_\gamma + c\sigma^2{}_\beta]} \cdot \frac{ab(c - 1)\sigma^2{}_\gamma}{\sum_{ijk} Y^2{}_{ijk} - \sum_{ij} Y^2{}_{ij.}/c} \sim F[a(b - 1), ab(c - 1)]$$

and when H_β is true $(\sigma^2{}_\beta = 0)$, this statistic reduces to

$$\frac{\sum_{ij} Y^2{}_{ij.}/c - \sum_{i} Y^2{}_{i..}/bc}{a(b - 1)} \cdot \frac{ab(c - 1)}{\sum_{ijk} Y^2{}_{ijk} - \sum_{ij} Y^2{}_{ij.}/c} = \frac{MS(B\text{'s in } A)}{MS(C\text{'s in } B)}$$

7.3 AN ILLUSTRATIVE TWO-FOLD HIERARCHAL PROBLEM

The analysis of two-fold hierarchal data will be illustrated by considering data related to plants on benches in greenhouses. Suppose that the objective is to determine the magnitudes of the variability associated with each source and then to decide if the small components of variance are significantly different from zero.

Let H_i denote the ith greenhouse and let B_{ij} denote the jth bench in the ith greenhouse. Consider an observational value Y_{ijk} for plant C_{ijk}.

H_1		H_2		H_3	
B_{11}	B_{12}	B_{21}	B_{22}	B_{31}	B_{32}
2.5	1.2	.9	2.7	5.0	4.0
3.0	2.7	1.2	2.0	4.2	4.0
2.9	3.0	1.3	2.1	4.1	3.6
8.4	6.9	3.4	6.8	13.3	11.6

$$Y_{1..} = 15.3 \qquad Y_{2..} = 10.2 \qquad Y_{3..} = 24.9$$
$$Y_{...} = 50.4$$

Assume now that $Y_{ijk} = \mu + \alpha_i + \beta_{ij} + \gamma_{ijk}$ where $\alpha_i \sim N(0, \sigma^2{}_\alpha)$, $\beta_{ij} \sim N(0, \sigma^2{}_\beta)$, $\gamma_{ijk} \sim N(0, \sigma^2{}_\gamma)$, and all random variables are independent, then an AOV can be written.

Source of variation	df	SS	MS	$\mathcal{E}MS$
Total corrected				
Among all plants	17	24.320		
Among greenhouses	2	18.570	9.285	$\sigma^2{}_\gamma + 3\sigma^2{}_\beta + 6\sigma^2{}_\alpha$
Between benches in greenhouses (pooled)	3	2.783	.928	$\sigma^2{}_\gamma + 3\sigma^2{}_\beta$
Among plants on benches (pooled)	12	2.967	.247	$\sigma^2{}_\gamma$

By solving the equations

$$\hat{\sigma}^2{}_\gamma + 3\hat{\sigma}^2{}_\beta + 6\hat{\sigma}^2{}_\alpha = 9.285$$
$$\hat{\sigma}^2{}_\gamma + 3\hat{\sigma}^2{}_\beta = .928$$
$$\hat{\sigma}^2{}_\gamma = .247$$

we obtain the unbiased estimates $\hat{\sigma}^2{}_\gamma = .247$, $\hat{\sigma}^2{}_\beta = .227$, and $\hat{\sigma}^2{}_\alpha = 1.393$. Thus it appears that the plant and bench components are about equal and the greenhouse component is about six times the magnitude of the other components. The computed F ratios

$$\frac{MS(\text{benches in greenhouses})}{MS(\text{among plants})} = 3.76$$

and

$$\frac{MS(\text{among greenhouses})}{MS(\text{benches in greenhouses})} = 10.0$$

are both significant at the .05 level.

7.4 EXERCISES

(1) Show that in the one-fold hierarchy

$$\mathcal{E}\left\{\left[\sum_i Y^2_{i.}/b - Y^2_{..}/ab\right]/(a-1)\right\} = \sigma^2_\beta + b\sigma^2_\alpha.$$

(2) For the two-fold hierarchal situation, determine the covariance of
   (a) Y_{ijk} and $Y_{ijk'}$ $k \neq k'$;
   (b) Y_{ijk} and $Y_{ij'k}$ $j \neq j'$;
   (c) Y_{ijk} and $Y_{i'jk}$ $i \neq i'$;
   (d) Y_{ijk} and $Y_{ij'k'}$ $j \neq j'$ and $k \neq k'$;
   (e) Y_{ijk} and $Y_{i'jk'}$ $i \neq i'$ and $k \neq k'$;
   (f) Y_{ijk} and $Y_{i'j'k}$ $i \neq i'$ and $j \neq j'$;
   (g) Y_{ijk} and $Y_{i'j'k'}$ $i \neq i'$, $j \neq j'$, and $k \neq k'$.

(3) Explain why the point estimates of variances such as σ^2_α are unbiased when we use the procedure suggested in the text (that is, when we solve the equations obtained by setting mean squares equal to expected mean squares).

(4) After a national election, the percent of voter turnout was determined in four precincts in each of three counties in four states. Assume a random-effects hierarchal model and analyze the variability by computing estimates of the components of variance.

State	A_1			A_2			A_3			A_4		
County	C_{11}	C_{12}	C_{13}	C_{21}	C_{22}	C_{23}	C_{31}	C_{32}	C_{33}	C_{41}	C_{42}	C_{43}
Percent	42	51	51	42	47	62	62	61	61	40	67	70
in precinct	47	49	50	41	51	61	59	60	59	47	59	71
	45	47	53	40	59	57	63	64	58	49	69	66
	42	46	49	42	60	60	64	59	57	49	70	58

(5)   A characteristic of a certain chemical showed much variability. In order to study this variability, three of many vats in which the chemical was produced were selected for the study. The daily production of the chemical was sampled on three days. The samples themselves were treated somewhat differently during and immediately before the measuring of the characteristic. Each measurement was taken twice.

Vat	A_1						A_2					
Day	B_{11}		B_{12}		B_{13}		B_{21}		B_{22}		B_{23}	
	C_{111}	C_{112}	C_{121}	C_{122}	C_{131}	C_{132}	C_{211}	C_{212}	C_{221}	C_{222}	C_{231}	C_{232}
	.72	.68	.59	.62	.47	.53	.69	.61	.55	.57	.81	.79
	.71	.68	.58	.63	.49	.52	.68	.61	.54	.57	.80	.79

Vat	A_3					
Day	B_{31}		B_{32}		B_{33}	
	C_{311}	C_{312}	C_{321}	C_{322}	C_{331}	C_{332}
	.47	.50	.62	.61	.45	.46
	.46	.51	.60	.59	.45	.47

Assume the random-effects hierarchal model

$$Y_{ijkl} = \mu + \alpha_i + \beta_{ij} + \gamma_{ijk} + \delta_{ijkl}$$

where $\alpha_i \sim N(0, \sigma^2_\alpha)$, $\beta_{ij} \sim N(0, \sigma^2_\beta)$, $\gamma_{ijk} \sim N(0, \sigma^2_\gamma)$, and $\delta_{ijkl} \sim N(0, \sigma^2_\delta)$. Estimate σ^2_α, σ^2_β, σ^2_γ, and σ^2_δ. Use the Snedecor F tables to determine which, if any, of the components of variance are significantly different from zero.

(6)   Assume the model $Y_{ij} = \mu + \tau_i + \epsilon_{ij}$ for the following data.

Treatment	I	II	III	IV	V
	1.7	1.8	1.3	1.8	1.4
	1.6	1.9	1.5	1.8	1.5
	1.5	2.0	1.2	2.0	1.6

Let $\tau_i \sim \text{NID}(0, \sigma^2_\tau)$, $\epsilon_{ij} \sim \text{NID}(0, \sigma^2_\epsilon)$ and assume that ϵ_{ij} and τ_i are independent. Determine an estimate of the variance of:
(a)   A new observation associated with a treatment picked at random.
(b)   The difference of two observations on experimental units with the same treatment.
(c)   The difference of two observations on experimental units with different treatments.

(d)   The average of five observations for experimental units with the same treatment (the treatment picked at random).

(e)   The average of any five observations for five randomly selected experimental units.

(7)   Consider the illustrative example of Section 7.3. Determine an estimate of the variance of:

   (a)   A new observation on a bench picked at random from a greenhouse picked at random.

   (b)   The average of five observations, all on the same bench in a randomly chosen greenhouse.

   (c)   The difference of two observations taken from the same bench in a greenhouse.

   (d)   The difference of two observations taken from different benches in the same greenhouse.

   (e)   The difference of two observations taken from different greenhouses.

7.5   *N*-WAY CROSS-CLASSIFICATION DESIGNS

In this section we wish to generalize the concepts presented when the frequently used two-way, cross-classification, fixed-effects model was developed. We have already seen that the model

$$Y_{ij} = \mu + \alpha_i + \beta_j + \epsilon_{ij}$$

is often appropriate in two-factor factorial experiments where the experimental units are known to be heterogeneous, and are grouped before randomization of treatments (a randomized block situation).

Consider the three-way cross-classification model

$$Y_{ijk} = \mu + \alpha_i + \beta_j + \gamma_k + \epsilon_{ijk}.$$

We have already discussed three-factor factorial experiments and their analysis. Such experiments are three-way cross-classification designs if the treatment combinations are assigned to experimental units at random. Other three-way cross-classification situations result when we:

(1)   Apply c treatments to abc heterogeneous experimental units where the units themselves have been cross-classified with c units in each of the ab categories of a two-way cross-classification.

(2)   Apply the bc treatment combinations of a two-factor experiment to the bc experimental units in each of "a" different blocks.

In the above situations and, indeed, in all n-way cross-classification designs, the analysis is made systematic by thinking of each classification as associated with a factor or pseudofactor. For example, we can think of

different blocks as different levels of a factor. All nonmeaningful interactions, as a general rule, are then pooled with experimental error. The
general cross-classification model is an extension of the three-way cross-
classification model.

The concepts will be illustrated in the following example. Suppose that
36 electrical experimental units were classified according to two electrical
characteristics (say voltage and amperage). Suppose four homogeneous
units were placed in each of the 3^2 factorial categories. These could be
referred to as nine blocks.

Voltage

Experimental units classified		High	Medium	Low
	High	(1, 1)	(1, 2)	(1, 3)
Amperage	Medium	(2, 1)	(2, 2)	(2, 3)
	Low	(3, 1)	(3, 2)	(3, 3)

Consider a 2^2 factorial arrangement of treatments associated with the
levels of factors C and D. Let the 2^2 treatment combinations be randomized over the four experimental units in each of the amperage-voltage
categories and let Y_{ijkl} denote the resulting observations.

Y_{ijkl} denotes the response of the (k, l) treatment on a unit earlier
classified as (i, j) with respect to amperage and voltage. Suppose that the
data were as follows:

Voltage

Amperage		High		Medium		Low		
	D level	C level		C level		C level		
		HIGH	LOW	HIGH	LOW	HIGH	LOW	TOTALS
	High	5.2	6.7	6.5	5.9	3.9	2.7	30.9
High	Low	8.7	7.5	6.6	6.2	4.5	2.8	36.3
	High	4.3	5.0	4.3	4.1	1.9	.9	20.5
Medium	Low	5.0	5.9	5.9	4.3	1.9	1.3	24.3
	High	2.3	5.0	3.8	3.0	1.1	.8	16.0
Low	Low	2.2	4.3	3.9	3.5	1.0	1.1	16.0
Totals		27.7	34.4	31.0	27.0	14.3	9.6	144.0

Assuming the fixed-effects model

$$Y_{ijkl} = \mu + \alpha_i + v_j + c_k + d_l + (cd)_{kl} + \epsilon_{ijkl}$$

the AOV is as follows:

Source	df	SS	MS	F
Total	36	723.72		
Mean	1	576.00		
Blocks	8	127.985	15.998	
C	1	.111	.111	
D	1	2.351	2.351	3.335
CD interaction	1	.360	.360	
Residual	24	16.913	.705	

Formally,

Blocks (8) $\equiv$ Amps(2) + Voltage(2) + AV interactions(4)

Residual (24)

$\equiv$ Block by treatment interactions (24)

$\equiv AC(2) + AD(2) + CV(2) + DV(2) + ACD(2) + VCD(2)$

$+ AVC(4) + AVD(4) + AVCD(4).$

The levels of D are declared significantly different at about the .08 level while the levels of C show little difference and the interaction appears very small. The AOV strongly supports the assumption that the experimental units were heterogeneous.

Perhaps the easiest way of obtaining the C, D, and CD sums of squares is to prepare a 2 × 2 table of totals.

In certain situations, a random-effects model is appropriate for cross-classification data. In other situations, the appropriate description seems to be *mixed-effects model*, where some components are fixed and others random. The exercises of the next section include situations where random-effects models and mixed-effects models are used to describe cross-classification data. Many of the computational details for analysis of data using these models are identical with the computational details associated with fixed-effects models. The principal differences are consequences of the fact that the expected mean squares for fixed, random, and mixed models are often considerably different. Consequently, tests of hypotheses can involve different ratios of mean squares as we vary the model from fixed to random to mixed. The expected mean squares for several situations appear in the exercises and the interested student is referred to one of the several good discussions found in the listed reference books.

7.6 EXERCISES

(1) A three-way cross-classification design was used to study the durability of superhighway surfaces. Three different climate areas were selected for study. Two depths were studied and four compositions were investigated. The data was collected for a number of repetitions of the design but the data for two repetitions appears here. Using differences between repetitions as a measure of experimental error, test for significance between depths, among compositions, and climate-composition interaction. Assume a fixed-effects model for your analysis.

		Composition							
		A_1		A_2		A_3		A_4	
DEPTH		D_1	D_2	D_1	D_2	D_1	D_2	D_1	D_2
	C_1	23.1	24.3	25.8	27.6	20.4	19.3	26.5	29.4
		23.7	25.4	21.9	26.9	20.3	19.8	26.1	29.8
Climate	C_2	20.7	28.9	24.3	26.9	19.4	24.1	19.5	24.6
		22.5	27.4	24.2	25.8	19.4	24.5	19.8	24.5
	C_3	20.5	19.6	25.4	24.9	21.7	22.5	20.5	20.7
		18.7	22.5	25.3	25.8	21.5	21.2	20.1	21.3

(2) Write out the sources of variation, including interactions, the degrees of freedom, and a fixed-effects model for an experiment where factor A is at three levels, factor B is at four levels, and the treatment combinations are run in five blocks where each block is composed of 24 homogeneous experimental units.

(3) Consider the random-effects model $Y_{ij} = \mu + \alpha_i + \beta_j + \epsilon_{ij}$ where μ is fixed but $\alpha_i \sim N(0, \sigma^2_\alpha)$, $\beta_j \sim N(0, \sigma^2_\beta)$, $\epsilon_{ij} \sim N(0, \sigma^2_\epsilon)$, and all random variables are independent. Work out the expected value of:

(a) $\displaystyle\sum_{i=1}^{a} \left(\frac{Y_{i.}}{b} - \frac{Y_{..}}{ab} \right)^2$

(b) $\displaystyle\sum_{j=1}^{b} \left(\frac{Y_{.j}}{a} - \frac{Y_{..}}{ab} \right)^2$

(4) An agronomy experiment was conducted on 18 Iowa farms which were selected more or less at random from all farms in the state. On each farm, four varieties of corn were planted in somewhat arbitrarily chosen test plots. A few years later the experiment was repeated on the same test plots,

and then repeated again the following year. The main objective was to study the variation in yields Y_{ijk} with respect to varieties, farms, and years. For this reason, and because previous studies had indicated small interactions, all of the interactions were ignored. The chosen model was

$$Y_{ijk} = \mu + \alpha_i + \beta_j + \gamma_k + e_{ijk}$$

where $\alpha_i \sim N(0, \sigma^2_\alpha)$, $\beta_j \sim N(0, \sigma^2_\beta)$, $\gamma_k \sim N(0, \sigma^2_\gamma)$, and $e_{ijk} \sim N(0, \sigma^2_e)$. All random variables were assumed to be independent. Complete the AOV and estimate σ^2_α, σ^2_β, σ^2_γ, and σ^2_e from the AOV.

Source	df	SS	MS	$\mathcal{E}MS$*
Total	216	51,736.1		
Mean	1	50,104.5		
Years (α)		306.8		$\sigma^2_e + bc\sigma^2_\alpha$
Farms (β)		353.6		$\sigma^2_e + ac\sigma^2_\beta$
Varieties (γ)		546.6		$\sigma^2_e + ab\sigma^2_\gamma$
Residual		424.6		σ^2_e

* In this column, a denotes the number of years, b the number of farms, and c the number of varieties.

(5) In a random-effects experiment similar to that described in Exercise (4), the objective was to test whether or not certain factors and interactions were significantly different from zero. In this experiment, each of the four varieties during each of the three years was replicated on each of the 18 farms in order to have a measure of experimental error. Complete the following AOV.

Source	df	SS	MS	$\mathcal{E}MS$*
Total	432	101,934.1		
Mean	1	99,546.2		
Years (α)		312.0		$\sigma^2_e + r\sigma^2_{\alpha\beta\gamma} + rc\sigma^2_{\alpha\beta} + rb\sigma^2_{\alpha\gamma} + rbc\sigma^2_\alpha$
Farms (β)		338.3		$\sigma^2_e + r\sigma^2_{\alpha\beta\gamma} + rc\sigma^2_{\alpha\beta} + ra\sigma^2_{\beta\gamma} + rac\sigma^2_\beta$
Varieties (γ)		566.4		$\sigma^2_e + r\sigma^2_{\alpha\beta\gamma} + rb\sigma^2_{\alpha\gamma} + ra\sigma^2_{\beta\gamma} + rab\sigma^2_\gamma$
$Y \times F$ interaction		346.8		$\sigma^2_e + r\sigma^2_{\alpha\beta\gamma} + rc\sigma^2_{\alpha\beta}$
$Y \times V$ interaction		243.6		$\sigma^2_e + r\sigma^2_{\alpha\beta\gamma} + rb\sigma^2_{\alpha\gamma}$
$F \times V$ interaction		163.2		$\sigma^2_e + r\sigma^2_{\alpha\beta\gamma} + ra\sigma^2_{\beta\gamma}$
$Y \times F \times V$		185.6		$\sigma^2_e + r\sigma^2_{\alpha\beta\gamma}$
Experimental error		232.0		σ^2_e

* The number r in the expected mean square column denotes the number of replications. In this experiment, $r = 2$.

By examining the expected mean squares determine which mean squares one would compare in order to test

(a) $\sigma^2_{\alpha\beta\gamma} = 0$

(b) $\sigma^2_{\beta\gamma} = 0$

(c) $\sigma^2_{\alpha\beta\gamma} = 0$ and $\sigma^2_{\alpha\beta} = 0$

(d) $\sigma^2_{\alpha\gamma} = 0$ and $\sigma^2_{\alpha} = 0$

(e) $\sigma^2_{\beta} = 0$

Do the data in Exercise (5) agree with interaction assumption made in connection with Exercise (4)? Do the findings in the two experiments tend to complement one another?

(6) Consider the mixed-effects model $Y_{ij} = \mu + \alpha_i + \beta_j + \epsilon_{ij}$ where μ, $\alpha_1, \ldots, \alpha_a$, are fixed, $\beta_j \sim N(0, \sigma^2_\beta)$, $j = 1, \ldots, b$, and $\epsilon_{ij} \sim N(0, \sigma^2_\epsilon)$. Work out the expected values of the following sums of squares under the assumption that all random variables are independent.

(a) $\displaystyle\sum_{i=1}^{a} (Y_{i.}/b - Y_{..}/ab)^2$

(b) $\displaystyle\sum_{j=1}^{b} (Y_{.j}/a + Y_{..}/ab)^2$

(7) The drying times for dresses of three different fabrics were recorded on four different days. Along with each drying time, the experimenter observed the temperature and found them to be very different. The data were analyzed, assuming a mixed-effects model $Y_{ijk} = \mu + \alpha_i + \beta_j + \gamma_{ij} + \epsilon_{ijk}$ where the fabric component α_i was assumed fixed, but the day (temperature) component $\beta_j \sim N(0, \sigma^2_\beta)$. γ_{ij} represented the random interaction component, and its distribution was assumed to be $N(0, \sigma^2_\gamma)$. Two dresses

Data

Day	A_1			A_2			A_3			A_4		
Fabric	B_1	B_2	B_3	B_1	B_2	B_3	B_1	B_2	B_3	B_1	B_2	B_3
Drying	2.18	1.14	1.51	1.22	1.31	1.01	2.85	2.04	1.96	.97	.87	1.24
times	2.30	1.20	1.48	1.23	1.27	.98	2.79	2.01	1.89	1.10	.85	1.30

AOV

Source	df	SS	MS	$\mathcal{E}MS$
Total	24			
Mean	1			
A fabrics	2			$\sigma^2_\epsilon + r\sigma^2_\gamma + \dfrac{rb}{a-1}\displaystyle\sum_{i=1}^{a}\alpha^2_i$
B days	3			$\sigma^2_\epsilon + r\sigma^2_\gamma + ra\sigma^2_\beta$
AB	6			$\sigma^2_\epsilon + r\sigma^2_\gamma$
Experimental error	12			σ^2_ϵ

of each fabric were dried starting at the same time each day. $\epsilon_{ij1} - \epsilon_{ij2} = Y_{ij1} - Y_{ij2}$ measures the differences between drying times for dresses treated alike. It was assumed that $\epsilon_{ijk} \sim N(0, \sigma^2_\epsilon)$. It was further assumed that all random components were independent.

(a) Complete the AOV.
(b) Test $\alpha_1 = \alpha_2 = \alpha_3$ (equality of fabric means) using a .05 probability for the Type I error.
(c) Estimate σ^2_β. (How would we test $\sigma^2_\beta = 0$?)
(d) Estimate σ^2_γ. (How would we test $\sigma^2_\gamma = 0$?)
(e) Verify the values of the expected mean squares recorded in the AOV.

7.7 ANALYSIS OF INCOMPLETE DATA

An experimental situation may be incomplete because it was designed to have incomplete classification or because observations were destroyed, lost, or impossible to obtain. A Latin square experiment is an example of an experiment where the data is, by design, incomplete. Formally, a $n \times n$ *Latin square* is a $n \times n$ array of n symbols where the ith symbol $i = 1, 2, \ldots, n$, appears once and only once in each row and each column. An example of a 4×4 Latin square is:

$$
\begin{array}{cccc}
A & B & C & D \\
C & D & A & B \\
D & A & B & C \\
B & C & D & A
\end{array}
$$

The Latin square concept has been used by many different experimenters, in situations as varied as growing grains in Kansas to placing batteries in rocket ships. In order to illustrate the use of this concept, consider an agronomist faced with a known two-way gradient of fertility in his plant-testing area. Suppose that four treatments are to be studied to determine the effect of the treatment on a plant yield. A procedure for handling the gradient problem is to assign treatments to letters A, B, C, and D and then apply the treatments to the $16 = 4^2$ test plots as indicated in the accompanying diagram. There are, of course, many different Latin squares and in situations where essentially the same experiment is repeated, it may be important to select at random a new Latin square each time the basic experiment is repeated.

If we now consider rows as a source of variation, columns as a second source of variation, and treatments as a third source of variation, then abstractly the design is a $\frac{1}{4}$ replicate of a complete 4^3 factorial experiment. Three-fourths of the possible observations, of necessity, have been omitted. For example, B, C, and D were not applied to the best soil in Row I and Column I. The incompleteness was by design and, for

Best soil

Column	I	II	III	IV	
Row I	*A*	*B*	*C*	*D*	*Gradient in soil fertility* ↓
II	*C*	*D*	*A*	*B*	
III	*D*	*A*	*B*	*C*	
IV	*B*	*C*	*D*	*A*	

Gradient in soil fertility → *Poorest soil*

obvious reasons, the analysis is that of a Latin square experimental design. Before writing out a model and displaying an AOV, let us consider other incomplete classification situations.

Suppose that $t = 3$ treatments are to be studied in conjunction with experimental units that are heterogeneous, but naturally can be grouped in pairs of homogeneous units. For example, the experimental units might be the $2b$ lambs resulting from b sets of twin births. In such an experimental situation, it is natural to arrange treatments and pairs in the following manner.

Treatment	I	II	III
Pair I	Y_{11}	Y_{12}	—
II	—	Y_{22}	Y_{23}
III	Y_{31}	—	Y_{33}

Let f denote the ratio of the number of observations in the incomplete design to the number bt of cells in the corresponding complete block design. In this case $f = \frac{2}{3}$. Members of the first pair are assigned to treatment I or treatment II, but treatment III is not applied. Members of the second pair are assigned at random to II or III with I not applied and members of the third pair are assigned to I and III with II not applied. The entire assignment could be repeated using additional sets of three pairs of experimental units. The design is a special case of what is commonly referred to as a *balanced incomplete block design*. We shall see later that, although this design and the Latin square design each have a degree of symmetry, there does exist a fundamental difference which is reflected in the fact that Latin square analysis of variance is computationally easier than balanced incomplete block analysis of variance.

As a third illustration of an incomplete design, consider the following two-way classification statistical layout where unintentionally a cell is empty. In this case the missing observation destroys most of the sym-

Treatment		I	II	III	IV
	I	Y_{11}	Y_{12}	Y_{13}	Y_{14}
Block	II	Y_{21}	Y_{22}	Y_{23}	Y_{24}
	III	Y_{31}	Y_{32}	Y_{33}	*Missing*

metrical features of the complete randomized block analysis of variance. There exist missing observation formulas devised to restore the symmetry and thereby restore the ease of hand calculating the required sums of squares for the analysis of variance. The interested student should see a reference for application of these methods.

Let us now return to the Latin square illustration. For this design, we construct a fixed-effects model and then, based on the model, we find point estimates of the differences between treatment population means, discuss the associated AOV, and test the hypothesis of equality of treatment means. If we rename the treatments 1, 2, 3, and 4 instead of A, B, C, and D, then for treatment k applied to the (i, j) experimental unit, let the response Y_{ijk} be written

$$Y_{ijk} = \mu + \alpha_i + \beta_j + \tau_k + \epsilon_{ijk}.$$

Here α_i denotes the effect of the ith row (in our illustration, the soil fertility in the plots of row i in the field layout), β_j denotes the effect of the jth column, τ_k denotes the effect of treatment k, leaving ϵ_{ijk} to represent all residual interaction and experimental error variation for the plot (i, j) with treatment k applied.

It is perhaps instructive to display the statistical layout in two forms.

Column		I	II	III	IV	Totals
	I	Y_{111}	Y_{122}	Y_{133}	Y_{144}	$Y_{1..}$
Row	II	Y_{213}	Y_{224}	Y_{231}	Y_{242}	$Y_{2..}$
	III	Y_{314}	Y_{321}	Y_{332}	Y_{343}	$Y_{3..}$
	IV	Y_{412}	Y_{423}	Y_{434}	Y_{441}	$Y_{4..}$
Totals		$Y_{.1.}$	$Y_{.2.}$	$Y_{.3.}$	$Y_{.4.}$	$Y_{...}$

Treatments	$A \equiv (1)$	$B \equiv (2)$	$C \equiv (3)$	$D \equiv (4)$
	Y_{111}	Y_{122}	Y_{133}	Y_{144}
	Y_{231}	Y_{242}	Y_{213}	Y_{224}
	Y_{321}	Y_{332}	Y_{343}	Y_{314}
	Y_{441}	Y_{412}	Y_{423}	Y_{434}
Totals	$Y_{..1}$	$Y_{..2}$	$Y_{..3}$	$Y_{..4}$

If we now assume that $\epsilon_{ijk} \sim \text{NID}(0, \sigma^2)$, then an unbiased estimate of the difference $\tau_k - \tau_l$ of the effects of treatment k and treatment l is

$$\hat{\tau}_k - \hat{\tau}_l = \frac{Y_{..k} - Y_{..l}}{r}$$

where r is the size of the Latin square. (In our example, $r = 4$.) Note that sums such as $Y_{..k}$ involve r observations, although the notation indicates that the summing operation has been applied to two indices. Likewise $Y_{...}$ is the sum of r^2 observations, although the notation indicates the summing operation for three indices.

The conventional AOV consists of partitioning the total sum of squares as displayed (r_0 denotes $r/r - 1$).

Source	df	SS	$\mathcal{E}MS$ (fixed effects)
Total	r^2	$\sum_{ijk} Y^2_{ijk}$	
Mean	1	$Y^2_{...}/r^2$	
Rows	$r-1$	$\sum_{i=1}^{r} Y^2_{i..}/r - Y^2_{...}/r^2$	$\sigma^2 + r_0\Sigma^2\alpha_i$
Columns	$r-1$	$\sum_{j=1}^{r} Y^2_{.j.}/r - Y^2_{...}/r^2$	$\sigma^2 + r_0\Sigma\beta^2_j$
Treatments	$r-1$	$\sum_{k=1}^{r} Y^2_{..k}/r - Y^2_{...}/r^2$	$\sigma^2 + r_0\Sigma\tau^2_k$
Residual	$(r-2)(r-1)$	By subtraction	σ^2

The expected mean squares follow readily from the assumption that $\epsilon_{ijk} \sim \text{NID}(0, \sigma^2)$. In displaying these expected mean squares, we have applied the usual restrictions, such as $\Sigma\alpha_i = 0$, in order to eliminate the overparameterization in the model. Once again, ratios of mean squares have Snedecor distributions under null hypotheses, $\Sigma\alpha^2_i = 0$, $\Sigma\beta^2_j = 0$, and $\Sigma\tau^2_k = 0$ so that, in order to test these hypotheses, we compare the computed F ratio with the tabulated upper-tail F ratio.

The expected values (for the same mean squares), under the assumption of random effects, are obtained by replacing $r_0\Sigma\alpha^2_i$ by $r\sigma^2_\alpha$, and so on.

Unfortunately, the situation is not so simple when we try to explain balanced incomplete block data with a fixed-effects model. Suppose that, for our earlier illustration, we let Y_{ij} represent the response when treatment j is applied to an experimental unit in the ith pair. Let $Y_{ij} = \mu +$

$\beta_i + \tau_j + \epsilon_{ij}$, $i = 1, 2, \ldots, b$, and $j = 1, 2, \ldots, t$. For our illustration

$$Y_{.2} - Y_{.1} = \beta_2 - \beta_3 + 2(\tau_1 - \tau_2) + \epsilon_{12} + \epsilon_{22} - \epsilon_{11} - \epsilon_{31}.$$

The expected value of $(Y_{.2} - Y_{.1})/fb$ is $\tau_1 - \tau_2 + \frac{1}{2}(\beta_2 - \beta_3)$. This, of course, means that when the fixed-effects model is used to describe the data, $\mathcal{E}[(Y_{.2} - Y_{.1})/fb]$ involves block differences as well as treatment differences. One way of eliminating the block difference from the expected value of the difference of treatment averages (or totals) is to assume a mixed-effects model with the block effects random and having expectation zero. Of course, this does not in itself justify the use of a mixed-effects model but, if on other grounds the mixed model can be justified, then its use does simplify the estimation problem. The details of showing that $(Y_{.j} - Y_{.k})/fb$ is an unbiased estimator of $\tau_j - \tau_k$ when we assume a mixed model are left as an exercise.

Balanced incomplete block experiments are typical of a large class of designs which, in general, are not analyzed by the methods presented in this book. The basic complicating factor is the fact mentioned above; that is, differences in treatment totals do not estimate treatment differences. This, in turn, complicates the AOV because in such cases block sum of squares and treatment sum of squares do not add to give the block and treatment sums of squares. These sums of squares are no longer orthogonal. Put in still another way, these sums of squares, when considered as random variables, are dependent. Fortunately for mathematical statisticians, the problems involved, both from theoretical and applied points of view, are handled more easily in matrix notation than in the summation notation of this text.

7.8 EXERCISES

(1) Write out three different 4×4 Latin squares.

(2) Picture the 27 points (i, j, k) in three-dimensional space when i, j, and k take on values 1, 2, and 3. From these 27 points, pick out the nine points associated with the following 3×3 Latin square:

$$
\begin{array}{ccc}
A & B & C \\
C & A & B \\
B & C & A
\end{array}
$$

where i denotes the row, j denotes the column, and we associate $k = 1$ with A, $k = 2$ with B, and $k = 3$ with C.

(3) Picture the nine points (i, j) in two-dimensional space associated with the balanced incomplete block discussed in Section 7.7. Indicate in the diagram the six points for which observations were obtained.

(4)   Show that if the fixed-effects model

$$Y_{ijk} = \mu + \alpha_i + \beta_j + \tau_k + \epsilon_{ijk}$$

is assumed for $(r \times r)$ Latin square data, then in the notation of Section 7.7, the random variable $(Y_{..k} - Y_{..l})/r$ is an unbiased estimator of $\tau_k - \tau_l$.

(5)   Write out the details for showing that $\mathcal{E}$ [treatment mean square] equals

$$\sigma^2 + r_0 \sum_{k=1}^{r} \tau^2_k$$ when the fixed-effects model $Y_{ijk} = \mu + \alpha_i + \beta_j + \tau_k + \epsilon_{ijk}$

is assumed for Latin square data.

(6)   Write out the details for showing that $\mathcal{E}$ [treatment mean square] equals $\sigma^2 + r\sigma^2_\tau$ when the random-effects model $Y_{ijk} = \mu + \alpha_i + \beta_j + \tau_k + \epsilon_{ijk}$ is assumed for Latin square data.

(7)   Sixteen experimental units were classified according to characteristics X, Y, and Z. Each classification had four levels, and the classification was done so that the data resulting from a planned experiment could be analyzed as a Latin square. Thus a unit classified as (2, 4, 1) would possess the second level of X, the fourth level of Y, and the first level of Z. Assume a random-effects model. Write out the AOV and estimate the components of variance, σ_X^2, σ_Y^2, and σ_Z^2.

(114) 9.3	(122) 5.1	(133) 5.2	(141) 8.7
(213) 8.7	(221) 6.0	(232) 4.1	(244) 8.5
(312) 7.6	(324) 9.8	(331) 9.9	(343) 6.2
(411) 9.0	(423) 4.2	(434) 8.3	(442) 5.1

APPENDIX

Table I The Standard Normal Distribution*

Entry is $p = P[Z < z]$

z	.00	.01	.02	.03	.04	.05	.06	.07	.08	.09
−.0	.5000	.4960	.4920	.4880	.4840	.4801	.4761	.4721	.4681	.4641
−.1	.4602	.4562	.4522	.4483	.4443	.4404	.4364	.4325	.4286	.4247
−.2	.4207	.4168	.4129	.4090	.4052	.4013	.3974	.3936	.3897	.3859
−.3	.3821	.3783	.3745	.3707	.3669	.3632	.3594	.3557	.3520	.3483
−.4	.3446	.3409	.3372	.3336	.3300	.3264	.3228	.3192	.3156	.3121
−.5	.3085	.3050	.3015	.2981	.2946	.2912	.2877	.2843	.2810	.2776
−.6	.2743	.2709	.2676	.2643	.2611	.2578	.2546	.2514	.2483	.2451
−.7	.2420	.2389	.2358	.2327	.2297	.2266	.2236	.2206	.2177	.2148
−.8	.2119	.2090	.2061	.2063	.2005	.1977	.1949	.1922	.1894	.1867
−.9	.1841	.1814	.1788	.1762	.1736	.1711	.1685	.1660	.1635	.1611
−1.0	.1587	.1562	.1539	.1515	.1492	.1469	.1446	.1423	.1401	.1379
−1.1	.1357	.1335	.1314	.1292	.1271	.1251	.1230	.1210	.1190	.1170
−1.2	.1151	.1131	.1112	.1093	.1075	.1056	.1038	.1020	.1003	.09853
−1.3	.09680	.09510	.09342	.09176	.09012	.08851	.08691	.08534	.08379	.08226
−1.4	.08076	.07927	.07780	.07636	.07493	.07353	.07215	.07078	.06944	.06811
−1.5	.06681	.06552	.06426	.06301	.06178	.06057	.05938	.05821	.05705	.05592
−1.6	.05480	.05370	.05262	.05155	.05050	.04947	.04846	.04746	.04648	.04551
−1.7	.04457	.04363	.04272	.04182	.04093	.04006	.03920	.03836	.03754	.03673
−1.8	.03593	.03515	.03438	.03362	.03288	.03216	.03144	.03074	.03005	.02938
−1.9	.02872	.02807	.02743	.02680	.02619	.02559	.02500	.02442	.02385	.02330

Table I (*Continued*)

z	.00	.01	.02	.03	.04	.05	.06	.07	.08	.09
−2.0	.02275	.02222	.02169	.02118	.02068	.02018	.01970	.01923	.01876	.01831
−2.1	.01786	.01743	.01700	.01659	.01616	.01578	.01539	.01500	.01463	.01426
−2.2	.01390	.01355	.01321	.01287	.01255	.01222	.01191	.01160	.01130	.01101
−2.3	.01072	.01044	.01017	$.0^29903$	$.0^29642$	$.0^29387$	$.0^29137$	$.0^28894$	$.0^28656$	$.0^28424$
−2.4	$.0^28198$	$.0^27976$	$.0^27760$	$.0^27549$	$.0^27344$	$.0^27143$	$.0^26947$	$.0^26756$	$.0^26569$	$.0^26387$
−2.5	$.0^26210$	$.0^26037$	$.0^25868$	$.0^25703$	$.0^25543$	$.0^25386$	$.0^25234$	$.0^25085$	$.0^24940$	$.0^24799$
−2.6	$.0^24661$	$.0^24527$	$.0^24396$	$.0^24269$	$.0^24145$	$.0^24025$	$.0^23907$	$.0^23793$	$.0^23681$	$.0^23573$
−2.7	$.0^23467$	$.0^23364$	$.0^23264$	$.0^23167$	$.0^23072$	$.0^22980$	$.0^22890$	$.0^22803$	$.0^22718$	$.0^22635$
−2.8	$.0^22555$	$.0^22477$	$.0^22401$	$.0^22327$	$.0^22256$	$.0^22186$	$.0^22118$	$.0^22052$	$.0^21988$	$.0^21926$
−2.9	$.0^21866$	$.0^21807$	$.0^21750$	$.0^21695$	$.0^21641$	$.0^21589$	$.0^21538$	$.0^21489$	$.0^21441$	$.0^21395$
−3.0	$.0^21350$	$.0^21306$	$.0^21264$	$.0^21223$	$.0^21183$	$.0^21114$	$.0^21107$	$.0^21070$	$.0^21035$	$.0^21001$
.0	.5000	.5040	.5080	.5120	.5160	.5199	.5239	.5279	.5319	.5359
.1	.5398	.5438	.5478	.5517	.5557	.5596	.5636	.5675	.5714	.5753
.2	.5793	.5832	.5871	.5910	.5948	.5987	.6026	.6064	.6103	.6141
.3	.6179	.6217	.6255	.6293	.6331	.6368	.6406	.6443	.6480	.6517
.4	.6554	.6591	.6628	.6664	.6700	.6736	.6772	.6808	.6844	.6879
.5	.6915	.6950	.6985	.7019	.7054	.7088	.7123	.7157	.7190	.7224
.6	.7257	.7291	.7324	.7357	.7389	.7422	.7454	.7486	.7517	.7549
.7	.7580	.7611	.7642	.7673	.7703	.7734	.7764	.7794	.7823	.7852
.8	.7881	.7910	.7939	.7967	.7995	.8023	.8051	.8078	.8106	.8133
.9	.8159	.8186	.8212	.8238	.8264	.8289	.8315	.8340	.8365	.8389

* Reprinted with permission from A. Hald, *Statistical Tables and Formulas*, John Wiley & Sons, Inc., New York, 1952.

1.0	.8413	.8438	.8461	.8485	.8508	.8531	.8554	.8577	.8599	.8621
1.1	.8643	.8665	.8686	.8708	.8729	.8749	.8770	.8790	.8810	.8830
1.2	.8849	.8869	.8888	.8907	.8925	.8944	.8962	.8980	.8997	.90147
1.3	.90320	.90490	.90658	.90824	.90988	.91149	.91309	.91466	.91621	.91774
1.4	.91924	.92073	.92220	.92364	.92507	.92647	.92785	.92922	.93056	.93189
1.5	.93319	.93448	.93574	.93669	.93822	.93943	.94062	.94179	.94295	.94408
1.6	.94520	.94630	.94738	.94845	.94950	.95053	.95154	.95254	.95352	.95449
1.7	.95543	.95637	.95728	.95818	.95907	.95994	.96080	.96164	.96246	.96327
1.8	.96407	.96485	.96562	.96638	.96712	.96784	.96856	.96926	.96995	.97062
1.9	.97128	.97193	.97257	.97320	.97381	.97441	.97500	.97558	.97615	.97670
2.0	.97725	.97778	.97831	.97882	.97932	.97982	.98030	.98077	.98124	.98169
2.1	.98214	.98257	.98300	.98341	.98382	.98422	.98461	.98500	.98537	.98574
2.2	.98610	.98645	.98679	.98713	.98745	.98778	.98809	.98840	.98870	.98899
2.3	.98928	.98956	.98983	$.9^2 0097$	$.9^2 0358$	$.9^2 0613$	$.9^2 0863$	$.9^2 1106$	$.9^2 1344$	$.9^2 1576$
2.4	$.9^2 1802$	$.9^2 2024$	$.9^2 2240$	$.9^2 2451$	$.9^2 2656$	$.9^2 2857$	$.9^2 3053$	$.9^2 3244$	$.9^2 3431$	$.9^2 3613$
2.5	$.9^2 3790$	$.9^2 3963$	$.9^2 4132$	$.9^2 4297$	$.9^2 4457$	$.9^2 4614$	$.9^2 4766$	$.9^2 4915$	$.9^2 5060$	$.9^2 5201$
2.6	$.9^2 5339$	$.9^2 5473$	$.9^2 5604$	$.9^2 5731$	$.9^2 5855$	$.9^2 5975$	$.9^2 6093$	$.9^2 6207$	$.9^2 6319$	$.9^2 6427$
2.7	$.9^2 6533$	$.9^2 6636$	$.9^2 6736$	$.9^2 6833$	$.9^2 6928$	$.9^2 7020$	$.9^2 7110$	$.9^2 7197$	$.9^2 7282$	$.9^2 7365$
2.8	$.9^2 7445$	$.9^2 7523$	$.9^2 7599$	$.9^2 7673$	$.9^2 7744$	$.9^2 7814$	$.9^2 7882$	$.9^2 7948$	$.9^2 8012$	$.9^2 8074$
2.9	$.9^2 8134$	$.9^2 8193$	$.9^2 8250$	$.9^2 8305$	$.9^2 8359$	$.9^2 8411$	$.9^2 8462$	$.9^2 8511$	$.9^2 8559$	$.9^2 8605$
3.0	$.9^2 8650$	$.9^2 8694$	$.9^2 8736$	$.9^2 8777$	$.9^2 8817$	$.9^2 8856$	$.9^2 8893$	$.9^2 8930$	$.9^2 8965$	$.9^2 8999$

Note: $.0^2 1350 = .001350$, $.9^2 8650 = .998650$.

Table II   The Chi Square Distribution*

Entry is ω where $P[W < \omega] = p$ and ν is the Degree of Freedom

ν \ p	.005	.010	.025	.050	.100	.900	.950	.975	.990	.995
1	0.0^4393	0.0^3157	0.0^3982	0.0^2393	0.0158	2.71	3.84	5.02	6.63	7.88
2	0.0100	0.0201	0.0506	0.103	0.211	4.61	5.99	7.38	9.21	10.60
3	0.072	0.115	0.216	0.352	0.584	6.25	7.81	7.81	11.34	12.84
4	0.207	0.297	0.484	0.711	1.064	7.78	9.49	11.14	13.28	14.86
5	0.412	0.554	0.831	1.145	1.61	9.24	11.07	12.83	15.09	16.75
6	0.676	0.872	1.24	1.64	2.20	10.64	12.59	14.45	16.81	18.55
7	0.989	1.24	1.69	2.17	2.83	12.02	14.07	16.01	18.48	20.28
8	1.34	1.65	2.18	2.73	3.49	13.36	15.51	17.53	20.09	21.96
9	1.73	2.09	2.70	3.33	4.17	14.68	16.92	19.02	21.67	23.59
10	2.16	2.56	3.25	3.94	4.87	15.99	18.31	20.48	23.21	25.19
11	2.60	3.05	3.82	4.57	5.58	17.28	19.68	21.92	24.72	26.76
12	3.07	3.57	4.40	5.23	6.30	18.55	21.03	23.34	26.22	28.30
13	3.57	4.11	5.01	5.89	7.04	19.81	22.36	24.74	27.69	29.82
14	4.07	4.66	5.63	6.57	7.79	21.06	23.68	26.12	29.14	31.32
15	4.60	5.23	6.26	7.26	8.55	22.31	25.00	27.49	30.58	32.80
16	5.14	5.81	6.91	7.96	9.31	23.54	26.30	28.85	32.00	34.27
17	5.70	6.41	7.56	8.67	10.09	24.77	27.59	30.19	33.41	35.72
18	6.26	7.01	8.23	9.39	10.86	25.99	28.87	31.53	34.81	37.16
19	6.84	7.63	8.91	10.12	11.65	27.20	30.14	32.85	36.19	38.58
20	7.43	8.26	8.59	10.85	12.44	28.41	31.41	34.17	37.57	40.00
21	8.03	8.90	10.28	11.59	13.24	29.62	32.67	35.48	38.93	41.40
22	8.64	9.54	10.98	12.34	14.04	30.81	33.92	36.78	40.29	42.80
23	9.26	10.20	11.69	13.09	14.85	32.01	35.17	38.08	41.64	44.18
24	9.89	10.86	12.40	13.85	15.66	33.20	36.42	39.36	42.98	45.56
25	10.52	11.52	13.12	14.61	16.47	34.38	37.65	40.65	44.31	46.93
26	11.16	12.20	13.84	15.38	17.29	35.56	38.89	41.92	45.64	48.29
27	11.81	12.88	14.57	16.15	18.11	36.74	40.11	43.19	46.96	49.64
28	12.46	13.56	15.31	16.93	18.94	37.92	41.34	44.46	48.28	50.99
29	13.21	14.26	16.05	17.71	19.77	39.09	42.56	45.72	49.59	52.34
30	13.79	14.95	16.79	18.49	20.60	40.26	43.77	46.98	50.89	53.67
40	20.71	22.16	24.43	26.51	29.05	51.80	55.76	59.34	63.69	66.77
50	27.99	29.71	32.36	34.76	37.69	63.17	67.50	71.42	76.15	79.49
60	35.53	37.48	40.48	43.19	46.46	74.40	79.08	83.30	88.38	91.95
70	43.28	45.44	48.76	51.74	55.33	85.53	90.53	95.02	100.4	104.2
80	51.17	53.54	57.15	60.39	64.28	96.58	101.9	106.6	112.3	116.3
90	59.20	61.75	65.65	69.13	73.29	107.6	113.1	118.1	124.1	128.3
100	67.33	70.06	74.22	77.93	82.36	118.5	124.3	129.6	135.8	140.2

* Extracted with permission from H. L. Harter, A New Table of Percentage Points of the Chi-square Distribution, *Biometrika*, June, 1964.

Table III The Student t Distribution*

Entry is t where $P[T < t] = p$ and ν is the Degree of Freedom

ν \ p	.60	.70	.80	.90	.95	.975	.990	.995	.999	.9995
1	.325	.727	1.376	3.078	6.314	12.71	31.82	63.66	318.3	636.6
2	.289	.617	1.061	1.886	2.920	4.303	6.965	9.925	22.33	31.60
3	.277	.584	.978	1.638	2.353	3.182	4.541	5.841	10.22	12.94
4	.271	.569	.941	1.533	2.132	2.776	3.747	4.604	7.173	8.610
5	.267	.559	.920	1.476	2.015	2.571	3.365	4.032	5.893	6.859
6	.265	.553	.906	1.440	1.943	2.447	3.143	3.707	5.208	5.959
7	.263	.549	.896	1.415	1.895	2.365	2.998	3.499	4.785	5.405
8	.262	.546	.889	1.397	1.860	2.306	2.896	3.355	4.501	5.041
9	.261	.543	.883	1.383	1.833	2.262	2.821	3.250	4.297	4.781
10	.260	.542	.879	1.372	1.812	2.228	2.764	3.169	4.144	4.587
11	.260	.540	.876	1.363	1.796	2.201	2.718	3.106	4.025	4.437
12	.259	.539	.873	1.356	1.782	2.179	2.681	3.055	3.930	4.318
13	.259	.538	.870	1.350	1.771	2.160	2.650	3.012	3.852	4.221
14	.258	.537	.868	1.345	1.761	2.145	2.624	2.977	3.787	4.140
15	.258	.536	.866	1.341	1.753	2.131	2.602	2.947	3.733	4.073
16	.258	.535	.865	1.337	1.746	2.120	2.583	2.921	3.686	4.015
17	.257	.534	.863	1.333	1.740	2.110	2.567	2.898	3.646	3.965
18	.257	.534	.862	1.330	1.734	2.101	2.552	2.878	3.611	3.922
19	.257	.533	.861	1.328	1.729	2.093	2.539	2.861	3.579	3.883
20	.257	.533	.860	.1325	1.725	2.086	2.528	2.845	3.552	3.850
21	.257	.532	.859	1.323	1.721	2.080	2.518	2.831	3.527	3.819
22	.256	.532	.858	1.321	1.717	2.074	2.508	2.819	3.505	3.792
23	.256	.532	.858	1.319	1.714	2.069	2.500	2.807	3.485	3.767
24	.256	.531	.857	1.318	1.711	2.064	2.492	2.797	3.467	3.745
25	.256	.531	.856	1.316	1.708	2.060	2.485	2.787	3.450	3.725
26	.256	.531	.856	1.315	1.706	2.056	2.479	2.779	3.435	3.707
27	.256	.531	.855	1.314	1.703	2.052	2.473	2.771	3.421	3.690
28	.256	.530	.855	1.313	1.701	2.048	2.467	2.763	3.408	3.674
29	.256	.530	.854	1.311	1.699	2.045	2.462	2.756	3.396	3.659
30	.256	.530	.854	1.310	1.697	2.042	2.457	2.750	3.385	3.646
40	.255	.529	.851	1.303	1.684	2.021	2.423	2.704	3.307	3.551
50	.255	.528	.849	1.298	1.676	2.009	2.403	2.678	3.262	3.495
60	.254	.527	.848	1.296	1.671	2.000	2.390	2.660	3.232	3.460
80	.254	.527	.846	1.292	1.664	1.990	2.374	2.639	3.195	3.415
100	.254	.526	.845	1.290	1.660	1.984	2.365	2.626	3.174	3.389
200	.254	.525	.843	1.286	1.653	1.972	2.345	2.601	3.131	3.339
500	.253	.525	.842	1.283	1.648	1.965	2.334	2.586	3.106	3.310
∞	.253	.524	.842	1.282	1.645	1.960	2.326	2.576	3.090	3.291

*Reprinted by permission from A. Hald, *Statistical Tables and Formulas* (John Wiley & Sons, Inc., New York, 1952). The majority of the entries are from Table III of R. A. Fisher and F. Yates, *Statistical Tables* (Oliver and Boyd, Edinburgh), and are reprinted by permission of the authors and publishers.

Table IV   The Snedecor F Distribution

Entry is f where $P[F > f] = p$

De-nomi-nator df	p	\multicolumn{9}{c}{Numerator df}								
		1	2	3	4	5	6	7	8	9
1	.100	39.86	49.50	53.59	55.83	57.24	58.20	58.91	59.44	59.86
	.050	161.4	199.5	215.7	224.6	230.2	234.0	236.8	238.9	240.5
	.025	647.8	799.5	864.2	899.6	921.8	937.1	948.2	956.7	963.3
	.010	4052	4999.5	5403	5625	5764	5859	5928	5982	6022
	.005	16211	20000	21615	22500	23056	23437	23715	23925	24091
2	.100	8.53	9.00	9.16	9.24	9.29	9.33	9.35	9.37	9.38
	.050	18.51	19.00	19.16	19.25	19.30	19.33	19.35	19.37	19.38
	.025	38.51	39.00	39.17	39.25	39.30	39.33	39.36	39.37	39.39
	.010	98.50	99.00	99.17	99.25	99.30	99.33	99.36	99.37	99.39
	.005	198.5	199.0	199.2	199.2	199.3	199.3	199.4	199.4	199.4
3	.100	5.54	5.46	5.39	5.34	5.31	5.28	5.27	5.25	5.24
	.050	10.13	9.55	9.28	9.12	9.01	8.94	8.89	8.85	8.81
	.025	17.44	16.04	15.44	15.10	14.88	14.73	14.62	14.54	14.47
	.010	34.12	30.82	29.46	28.71	28.24	27.91	27.67	27.49	37.35
	.005	55.55	49.80	47.47	46.19	45.39	44.84	44.43	44.13	43.88
4	.100	4.54	4.32	4.19	4.11	4.05	4.01	3.98	3.95	3.94
	.050	7.71	6.94	6.59	6.39	6.26	6.16	6.09	6.04	6.00
	.025	12.22	10.65	9.98	9.60	9.36	9.20	9.07	8.98	8.90
	.010	21.20	18.00	16.69	15.98	15.52	15.21	14.98	14.80	14.66
	.005	31.33	26.28	24.26	23.15	22.46	21.97	21.62	12.35	21.14
5	.100	4.06	3.78	3.62	3.52	3.45	3.40	3.37	3.34	3.32
	.050	6.61	5.79	5.41	5.19	5.05	4.95	4.88	4.82	4.77
	.025	10.01	8.43	7.76	7.39	7.15	6.98	6.85	6.76	6.68
	.010	16.26	13.27	12.06	11.39	10.97	10.67	10.46	10.29	10.16
	.005	22.78	18.31	16.53	15.56	14.94	14.51	14.20	13.96	13.77
6	.100	3.78	3.46	3.29	3.18	3.11	3.05	3.01	2.98	2.96
	.050	5.99	5.14	4.76	4.53	4.39	4.28	4.21	4.15	4.10
	.025	8.81	7.26	6.60	6.23	5.99	5.82	5.70	5.60	5.52
	.010	13.75	10.92	9.78	9.15	8.75	8.47	8.26	8.10	7.98
	.005	18.63	14.54	12.92	12.03	11.46	11.07	10.79	10.57	10.39
7	.100	3.59	3.26	3.07	2.96	2.88	2.83	2.78	2.75	2.72
	.050	5.59	4.74	4.35	4.12	3.97	3.87	3.79	3.73	3.68
	.025	8.07	6.54	5.89	5.52	5.29	5.12	4.99	4.90	4.82
	.010	12.25	9.55	8.45	7.85	7.46	7.19	6.99	6.84	6.72
	.005	16.24	12.40	10.88	10.05	9.52	9.16	8.89	8.68	8.51
8	.100	3.46	3.11	2.92	2.81	2.73	2.67	2.62	2.59	2.56
	.050	5.32	4.46	4.07	3.84	3.69	3.58	3.50	3.44	3.39
	.025	7.57	6.06	5.42	5.05	4.82	4.65	4.53	4.43	4.36
	.010	11.26	8.65	7.59	7.01	6.63	6.37	6.18	6.03	5.91
	.005	14.69	11.04	9.60	8.81	8.30	7.95	7.69	7.50	7.34
9	.100	3.36	3.01	2.81	2.69	2.61	2.55	2.51	2.47	2.44
	.050	5.12	4.26	3.86	3.63	3.48	3.37	3.29	3.23	3.18
	.025	7.21	5.71	5.08	4.72	4.48	4.32	4.20	4.10	4.03
	.010	10.56	8.02	6.99	6.42	6.06	5.80	5.61	5.47	5.35
	.005	13.61	10.11	8.72	7.96	7.47	7.13	6.88	6.69	6.54
10	.100	3.29	2.92	2.73	2.61	2.52	2.46	2.41	2.38	2.35
	.050	4.96	4.10	3.71	3.48	3.33	3.22	3.14	3.07	3.02
	.025	6.94	5.46	4.83	4.47	4.24	4.07	3.95	3.85	3.78
	.010	10.04	7.56	6.55	5.99	5.64	5.39	5.20	5.06	4.94
	.005	12.83	9.43	8.08	7.34	6.87	6.54	6.30	6.12	5.97

Table IV *(Continued)*

Numerator *df*

10	12	15	20	24	30	40	60	120	∞	p	df
60.19	60.71	61.22	61.74	62.00	62.26	62.53	62.79	63.06	63.33	.100	1
241.9	243.9	245.9	248.0	249.1	250.1	251.1	252.2	253.3	254.3	.050	
968.6	976.7	984.9	993.1	997.2	1001	1006	1010	1014	1018	.025	
6056	6106	6157	6209	6235	6261	6287	6313	6339	6366	.010	
24224	24426	24630	24836	24940	25044	25148	25253	25359	25465	.005	
9.39	9.41	9.42	9.44	9.45	9.46	9.47	9.47	9.48	9.49	.100	2
19.40	19.41	19.43	19.45	19.45	19.46	19.47	19.48	19.49	19.50	.050	
39.40	39.41	39.43	39.45	39.46	39.46	39.47	39.48	39.49	39.50	.025	
99.40	99.42	99.43	99.45	99.46	99.47	99.47	99.48	99.49	99.50	.010	
199.4	199.4	199.4	199.4	199.5	199.5	199.5	199.5	199.5	199.5	.005	
5.23	5.22	5.20	5.18	5.18	5.17	5.16	5.15	5.14	5.13	.100	3
8.79	8.74	8.70	8.66	8.64	8.62	8.59	8.57	8.55	8.53	.050	
14.42	14.34	14.25	14.17	14.12	14.08	14.04	13.99	13.95	13.90	.025	
27.23	27.05	26.87	26.69	26.60	26.50	26.41	26.32	26.22	26.13	.010	
43.69	43.39	43.08	42.78	42.62	42.47	42.31	42.15	41.99	41.83	.005	
3.92	3.90	3.87	3.84	3.83	3.82	3.80	3.79	3.78	3.76	.100	4
5.96	5.91	5.86	5.80	5.77	5.75	5.72	5.69	5.66	5.63	.050	
8.84	8.75	8.66	8.56	8.51	8.46	8.41	8.36	8.31	8.26	.025	
14.55	14.37	14.20	14.02	13.93	13.84	13.75	13.65	13.56	13.46	.010	
20.97	20.70	20.44	20.17	20.03	19.89	19.75	19.61	19.47	19.32	.005	
3.30	3.27	3.24	3.21	3.19	3.17	3.16	3.14	3.12	3.10	.100	5
4.74	4.68	4.62	4.56	4.53	4.50	4.46	4.43	4.40	4.36	.050	
6.62	6.52	6.43	6.33	6.28	6.23	6.18	6.12	6.07	6.02	.025	
10.05	9.89	9.72	9.55	9.47	9.38	9.29	9.20	9.11	9.02	.010	
13.62	13.38	13.15	12.90	12.78	12.66	12.53	12.40	12.27	12.14	.005	
2.94	2.90	2.87	2.84	2.82	2.80	2.78	2.76	2.74	2.72	.100	6
4.06	4.00	3.94	3.87	3.84	3.81	3.77	3.74	3.70	3.67	.050	
5.46	5.37	5.27	5.17	5.12	5.07	5.01	4.96	4.90	4.85	.025	
7.87	7.72	7.56	7.40	7.31	7.23	7.14	7.06	6.97	6.88	.010	
10.25	10.03	9.81	9.59	9.47	9.36	9.24	9.12	9.00	8.88	.005	
2.70	2.67	2.63	2.59	2.58	2.56	2.54	2.51	2.49	2.47	.100	7
3.64	3.57	3.51	3.44	3.41	3.38	3.34	3.30	3.27	3.23	.050	
4.76	4.67	4.57	4.47	4.42	4.36	4.31	4.25	4.20	4.14	.025	
6.62	6.47	6.31	6.16	6.07	5.99	5.91	5.82	5.74	5.65	.010	
8.38	8.18	7.97	7.75	7.65	7.53	7.42	7.31	7.19	7.08	.005	
2.54	2.50	2.46	2.42	2.40	2.38	2.36	2.34	2.32	2.29	.100	8
3.35	3.28	3.22	3.15	3.12	3.08	3.04	3.01	2.97	2.93	.050	
4.30	4.20	4.10	4.00	3.95	3.89	3.84	3.78	3.73	3.67	.025	
5.81	5.67	5.52	5.36	5.28	5.20	5.12	5.03	4.95	4.86	.010	
7.21	7.01	6.81	6.61	6.50	6.40	6.29	6.18	6.06	5.95	.005	
2.42	2.38	2.34	2.30	2.28	2.25	2.23	2.21	2.18	2.16	.100	9
3.14	3.07	3.01	2.94	2.90	2.86	2.83	2.79	2.75	2.71	.050	
3.96	3.87	3.77	3.67	3.61	3.56	3.51	3.45	3.39	3.33	.025	
5.26	5.11	4.96	4.81	4.73	4.65	4.57	4.48	4.40	4.31	.010	
6.42	6.23	6.03	5.83	5.73	5.62	5.52	5.41	5.30	5.19	.005	
2.32	2.28	2.24	2.20	2.18	2.16	2.13	2.11	2.08	2.06	.100	10
2.98	2.91	2.85	2.77	2.74	2.70	2.66	2.62	2.58	2.54	.050	
3.72	3.62	3.52	3.42	3.37	3.31	3.26	3.20	3.14	3.08	.025	
4.85	4.71	4.56	4.41	4.33	4.25	4.17	4.08	4.00	3.91	.010	
5.85	5.66	5.47	5.27	5.17	5.07	4.97	4.86	4.75	4.64	.005	

Table IV (*Continued*)

De-nomi-nator df	p	1	2	3	4	5	6	7	8	9
						Numerator df				
11	.100	3.23	2.86	2.66	2.54	2.45	2.39	2.34	2.30	2.27
	.050	4.84	3.98	3.59	3.36	3.20	3.09	3.01	2.95	2.90
	.025	6.72	5.26	4.63	4.28	4.04	3.88	3.76	3.66	3.59
	.010	9.65	7.21	6.22	5.67	5.32	5.07	4.89	4.74	4.63
	.005	12.23	8.91	7.60	6.88	6.42	6.10	5.86	5.68	5.54
12	.100	3.18	2.81	2.61	2.48	2.39	2.33	2.28	2.24	2.21
	.050	4.75	3.89	3.49	3.26	3.11	3.00	2.91	2.85	2.80
	.025	6.55	5.10	4.47	4.12	3.89	3.73	3.61	3.51	3.44
	.010	9.33	6.93	5.95	5.41	5.06	4.82	4.64	4.50	4.39
	.005	11.75	8.51	7.23	6.52	6.07	5.76	5.52	5.35	5.20
13	.100	3.14	2.76	2.56	2.43	2.35	2.28	2.23	2.20	2.16
	.050	4.67	3.81	3.41	3.18	3.03	2.92	2.83	2.77	2.71
	.025	6.41	4.97	4.35	4.00	3.77	3.60	3.48	3.39	3.31
	.010	9.07	6.70	5.74	5.21	4.86	4.62	4.44	4.30	4.19
	.005	11.37	8.19	6.93	6.23	5.79	5.48	5.25	5.08	4.94
14	.100	3.10	2.73	2.52	2.39	2.31	2.24	2.19	2.15	2.12
	.050	4.60	3.74	3.34	3.11	2.96	2.85	2.76	2.70	2.65
	.025	6.30	4.86	4.24	3.89	3.66	3.50	3.38	3.29	3.21
	.010	8.86	6.51	5.56	5.04	4.69	4.46	4.28	4.14	4.03
	.005	11.06	7.92	6.68	6.00	5.56	5.26	5.03	4.86	4.72
15	.100	3.07	2.70	2.49	2.36	2.27	2.21	2.16	2.12	2.09
	.050	4.54	3.68	3.29	3.06	2.90	2.79	2.71	2.64	2.59
	.025	6.20	4.77	4.15	3.80	3.58	3.41	3.29	3.20	3.12
	.010	8.68	6.36	5.42	4.89	4.56	4.32	4.14	4.00	3.89
	.005	10.80	7.70	6.48	5.80	5.37	5.07	4.85	4.67	4.54
16	.100	3.05	2.67	2.46	2.33	2.24	2.18	2.13	2.09	2.06
	.050	4.49	3.63	3.24	3.01	2.85	2.74	2.66	2.59	2.54
	.025	6.12	4.69	4.08	3.73	3.50	3.34	3.22	3.12	3.05
	.010	8.53	6.23	5.29	4.77	4.44	4.20	4.03	3.89	3.78
	.005	10.58	7.51	6.30	5.64	5.21	4.91	4.69	4.52	4.38
17	.100	3.03	2.64	2.44	2.31	2.22	2.15	2.10	2.06	2.03
	.050	4.45	3.59	3.20	2.96	2.81	2.70	2.61	2.55	2.49
	.025	6.04	4.62	4.01	3.66	3.44	3.28	3.16	3.06	2.98
	.010	8.40	6.11	5.18	4.67	4.34	4.10	3.93	3.79	3.68
	.005	10.38	7.35	6.16	5.50	5.07	4.78	4.56	4.39	4.25
18	.100	3.01	2.62	2.42	2.29	2.20	2.13	2.08	2.04	2.00
	.050	4.41	3.55	3.16	2.93	2.77	2.66	2.58	2.51	2.46
	.025	5.98	4.56	3.95	3.61	3.38	3.22	3.10	3.01	2.93
	.010	8.29	6.01	5.09	4.58	4.25	4.01	3.84	3.71	3.60
	.005	10.22	7.21	6.03	5.37	4.96	4.66	4.44	4.28	4.14
19	.100	2.99	2.61	2.40	2.27	2.18	2.11	2.06	2.02	1.98
	.050	4.38	3.52	3.13	2.90	2.74	2.63	2.54	2.48	2.42
	.025	5.92	4.51	3.90	3.56	3.33	3.17	3.05	2.96	2.88
	.010	8.18	5.93	5.01	4.50	4.17	3.94	3.77	3.63	3.52
	.005	10.07	7.09	5.92	5.27	4.85	4.56	4.34	4.18	4.04
20	.100	2.97	2.59	2.38	2.25	2.16	2.09	2.04	2.00	1.96
	.050	4.35	3.49	3.10	2.87	2.71	2.60	2.51	2.45	2.39
	.025	5.87	4.46	3.86	3.51	3.29	3.13	3.01	2.91	2.84
	.010	8.10	5.85	4.94	4.43	4.10	3.87	3.70	3.56	3.46
	.005	9.94	6.99	5.82	5.17	4.76	4.47	4.26	4.09	3.96

Table IV (*Continued*)

Numerator *df*

10	12	15	20	24	30	40	60	120	∞	*p*	*df*
2.25	2.21	2.17	2.12	2.10	2.08	2.05	2.03	2.00	1.97	.100	11
2.85	2.79	2.72	2.65	2.61	2.57	2.53	2.49	2.45	2.40	.050	
3.53	3.43	3.33	3.23	3.17	3.12	3.06	3.00	2.94	2.88	.025	
4.54	4.40	4.25	4.10	4.02	3.94	3.86	3.78	3.69	3.60	.010	
5.42	5.24	5.05	4.86	4.76	4.65	4.55	4.44	4.34	4.23	.005	
2.19	2.15	2.10	2.06	2.04	2.01	1.99	1.96	1.93	1.90	.100	12
2.75	2.69	2.62	2.54	2.51	2.47	2.43	2.38	2.34	2.30	.050	
3.37	3.28	3.18	3.07	3.02	2.96	2.91	2.85	2.79	2.72	.025	
4.30	4.16	4.01	3.86	3.78	3.70	3.62	3.54	3.45	3.36	.010	
5.09	4.91	4.72	4.53	4.43	4.33	4.23	4.12	4.01	3.90	.005	
2.14	2.10	2.05	2.01	1.98	1.96	1.93	1.90	1.88	1.85	.100	13
2.67	2.60	2.53	2.46	2.42	2.38	2.34	2.30	2.25	2.21	.050	
3.25	3.15	3.05	2.95	2.89	2.84	2.78	2.72	2.66	2.60	.025	
4.10	3.96	3.82	3.66	3.59	3.51	3.43	3.34	3.25	3.17	.010	
4.82	4.64	4.46	4.27	4.17	4.07	3.97	3.87	3.76	3.65	.005	
2.10	2.05	2.01	1.96	1.94	1.91	1.89	1.86	1.83	1.80	.100	14
2.60	2.53	2.46	2.39	2.35	2.31	2.27	2.22	2.18	2.13	.050	
3.15	3.05	2.95	2.84	2.79	2.73	2.67	2.61	2.55	2.49	.025	
3.94	3.80	3.66	3.51	3.43	3.35	3.27	3.18	3.09	3.00	.010	
4.60	4.43	4.25	4.06	3.96	3.86	3.76	3.66	3.55	3.44	.005	
2.06	2.02	1.97	1.92	1.90	1.87	1.85	1.82	1.79	1.76	.100	15
2.54	2.48	2.40	2.33	2.29	2.25	2.20	2.16	2.11	2.07	.050	
3.06	2.96	2.86	2.76	2.70	2.64	2.59	2.52	2.46	2.40	.025	
3.80	3.67	3.52	3.37	3.29	3.21	3.13	3.05	2.96	2.87	.010	
4.42	4.25	4.07	3.88	3.79	3.69	3.58	3.48	3.37	3.26	.005	
2.03	1.99	1.94	1.89	1.87	1.84	1.81	1.78	1.75	1.72	.100	16
2.49	2.42	2.35	2.28	2.24	2.19	2.15	2.11	2.06	2.01	.050	
2.99	2.89	2.79	2.68	2.63	2.57	2.51	2.45	2.38	2.32	.025	
3.69	3.55	3.41	3.26	3.18	3.10	3.02	2.93	2.84	2.75	.010	
4.27	4.10	3.92	3.73	3.64	3.54	3.44	3.33	3.22	3.11	.005	
2.00	1.96	1.91	1.86	1.84	1.81	1.78	1.75	1.72	1.69	.100	17
2.45	2.38	2.31	2.23	2.19	2.15	2.10	2.06	2.01	1.96	.050	
2.92	2.82	2.72	2.62	2.56	2.50	2.44	2.38	2.32	2.25	.025	
3.59	3.46	3.31	3.16	3.08	3.00	2.92	2.83	2.75	2.65	.010	
4.14	3.97	3.79	3.61	3.51	3.41	3.31	3.21	3.10	2.98	.005	
1.98	1.93	1.89	1.84	1.81	1.78	1.75	1.72	1.69	1.66	.100	18
2.41	2.34	2.27	2.19	2.15	2.11	2.06	2.02	1.97	1.92	.050	
2.87	2.77	2.67	2.56	2.50	2.44	2.38	2.32	2.26	2.19	.025	
3.51	3.37	3.23	3.08	3.00	2.92	2.84	2.75	2.66	2.57	.010	
4.03	3.86	3.68	3.50	3.40	3.30	3.20	3.10	2.99	2.87	.005	
1.96	1.91	1.86	1.81	1.79	1.76	1.73	1.70	1.67	1.63	.100	19
2.38	2.31	2.23	2.16	2.11	2.07	2.03	1.98	1.93	1.88	.050	
2.82	2.72	2.62	2.51	2.45	2.39	2.33	2.27	2.20	2.13	.025	
3.43	3.30	3.15	3.00	2.92	2.84	2.76	2.67	2.58	2.49	.010	
3.93	3.76	3.59	3.40	3.31	3.21	3.11	3.00	2.89	2.78	.005	
1.94	1.89	1.84	1.79	1.77	1.74	1.71	1.68	1.64	1.61	.100	20
2.35	2.28	2.20	2.12	2.08	2.04	1.99	1.95	1.90	1.84	.050	
2.77	2.68	2.57	2.46	2.41	2.35	2.29	2.22	2.16	2.09	.025	
3.37	3.23	3.09	2.94	2.86	2.78	2.69	2.61	2.52	2.42	.010	
3.85	3.68	3.50	3.32	3.22	3.12	3.02	2.92	2.81	2.69	.005	

Table IV (*Continued*)

De-nomi-nator df	p	Numerator df 1	2	3	4	5	6	7	8	9
21	.100	2.96	2.57	2.36	2.23	2.14	2.08	2.02	1.98	1.95
	.050	4.32	3.47	3.07	2.84	2.68	2.57	2.49	2.42	2.37
	.025	5.83	4.42	3.82	3.48	3.25	3.09	2.97	2.87	2.80
	.010	8.02	5.78	4.87	4.37	4.04	3.81	3.64	3.51	3.40
	.005	9.83	6.89	5.73	5.09	4.68	4.39	4.18	4.01	3.88
22	.100	2.95	2.56	2.35	2.22	2.13	2.06	2.01	1.97	1.93
	.050	4.30	3.44	3.05	2.82	2.66	2.55	2.46	2.40	2.34
	.025	5.79	4.38	3.78	3.44	3.22	3.05	2.93	2.84	2.76
	.010	7.95	5.72	4.82	4.31	3.99	3.76	3.59	3.45	3.35
	.005	9.73	6.81	5.65	5.02	4.61	4.32	4.11	3.94	3.81
23	.100	2.94	2.55	2.34	2.21	2.11	2.05	1.99	1.95	1.92
	.050	4.28	3.42	3.03	2.80	2.64	2.53	2.44	2.37	2.32
	.025	5.75	4.35	3.75	3.41	3.18	3.02	2.90	2.81	2.73
	.010	7.88	5.66	4.76	4.26	3.94	3.71	3.54	3.41	3.30
	.005	9.63	6.73	5.58	4.95	4.54	4.26	4.05	3.88	3.75
24	.100	2.93	2.54	2.33	2.19	2.10	2.04	1.98	1.94	1.91
	.050	4.26	3.40	3.01	2.78	2.62	2.51	2.42	2.36	2.30
	.025	5.72	4.32	3.72	3.38	3.15	2.99	2.87	2.78	2.70
	.010	7.82	5.61	4.72	4.22	3.90	3.67	3.50	3.36	3.26
	.005	9.55	6.66	5.52	4.89	4.49	4.20	3.99	3.83	3.69
25	.100	2.92	2.53	2.32	2.18	2.09	2.02	1.97	1.93	1.89
	.050	4.24	3.39	2.99	2.76	2.60	2.49	2.40	2.34	2.28
	.025	5.69	4.29	3.69	3.35	3.13	2.97	2.85	2.75	2.68
	.010	7.77	5.57	4.68	4.18	3.85	3.63	3.46	3.32	3.22
	.005	9.48	6.60	5.46	4.84	4.43	4.15	3.94	3.78	3.64
26	.100	2.91	2.52	2.31	2.17	2.08	2.01	1.96	1.92	1.88
	.050	4.23	3.37	2.98	2.74	2.59	2.47	2.39	2.32	2.27
	.025	5.66	4.27	3.67	3.33	3.10	2.94	2.82	2.73	2.65
	.010	7.72	5.53	4.64	4.14	3.82	3.59	3.42	3.29	3.18
	.005	9.41	6.54	5.41	4.79	4.38	4.10	3.89	3.73	3.60
27	.100	2.90	2.51	2.30	2.17	2.07	2.00	1.95	1.91	1.87
	.050	4.21	3.35	2.96	2.73	2.57	2.46	2.37	2.31	2.25
	.025	5.63	4.24	3.65	3.31	3.08	2.92	2.80	2.71	2.63
	.010	7.68	5.49	4.60	4.11	3.78	3.56	3.39	3.26	3.15
	.005	9.34	6.49	5.36	4.74	4.34	4.06	3.85	3.69	3.56
28	.100	2.89	2.50	2.29	2.16	2.06	2.00	1.94	1.90	1.87
	.050	4.20	3.34	2.95	2.71	2.56	2.45	2.36	2.29	2.24
	.025	5.61	4.22	3.63	3.29	3.06	2.90	2.78	2.69	2.61
	.010	7.64	5.45	4.57	4.07	3.75	3.53	3.36	3.23	3.12
	.005	9.28	6.44	5.32	4.70	4.30	4.02	3.81	3.65	3.52
29	.100	2.89	2.50	2.28	2.15	2.06	1.99	1.93	1.89	1.86
	.050	4.18	3.33	2.93	2.70	2.55	2.43	2.35	2.28	2.22
	.025	5.59	4.20	3.61	3.27	3.04	2.88	2.76	2.67	2.59
	.010	7.60	5.42	4.54	4.04	3.73	3.50	3.33	3.20	3.09
	.005	9.23	6.40	5.28	4.66	4.26	3.98	3.77	3.61	3.48
30	.100	2.88	2.49	2.28	2.14	2.05	1.98	1.93	1.88	1.85
	.050	4.17	3.32	2.92	2.69	2.53	2.42	2.33	2.27	2.21
	.025	5.57	4.18	3.59	3.25	3.03	2.87	2.75	2.65	2.57
	.010	7.56	5.39	4.51	4.02	3.70	3.47	3.30	3.17	3.07
	.005	9.18	6.35	5.24	4.62	4.23	3.95	3.74	3.58	3.45

Table IV *(Continued)*

			Numerator df								
10	12	15	20	24	30	40	60	120	∞	p	df
1.92	1.87	1.83	1.78	1.75	1.72	1.69	1.66	1.62	1.59	.100	21
2.32	2.25	2.18	2.10	2.05	2.01	1.96	1.92	1.87	1.81	.050	
2.73	2.64	2.53	2.42	2.37	2.31	2.25	2.18	2.11	2.04	.025	
3.31	3.17	3.03	2.88	2.80	2.72	2.64	2.55	2.46	2.36	.010	
3.77	3.60	3.43	3.24	3.15	3.05	2.95	2.84	2.73	2.61	.005	
1.90	1.86	1.81	1.76	1.73	1.70	1.67	1.64	1.60	1.57	.100	22
2.30	2.23	2.15	2.07	2.03	1.98	1.94	1.89	1.84	1.78	.050	
2.70	2.60	2.50	2.39	2.33	2.27	2.21	2.14	2.08	2.00	.025	
3.26	3.12	2.98	2.83	2.75	2.67	2.58	2.50	2.40	2.31	.010	
3.70	3.54	3.36	3.18	3.08	2.98	2.88	2.77	2.66	2.55	.005	
1.89	1.84	1.80	1.74	1.72	1.69	1.66	1.62	1.59	1.55	.100	23
2.27	2.20	2.13	2.05	2.01	1.96	1.91	1.86	1.81	1.76	.050	
2.67	2.57	2.47	2.36	2.30	2.24	2.18	2.11	2.04	1.97	.025	
3.21	3.07	2.93	2.78	2.70	2.62	2.54	2.45	2.35	2.26	.010	
3.64	3.47	3.30	3.12	3.02	2.92	2.82	2.71	2.60	2.48	.005	
1.88	1.83	1.78	1.73	1.70	1.67	1.64	1.61	1.57	1.53	.100	24
2.25	2.18	2.11	2.03	1.98	1.94	1.89	1.84	1.79	1.73	.050	
2.64	2.54	2.44	2.33	2.27	2.21	2.15	2.08	2.01	1.94	.025	
3.17	3.03	2.89	2.74	2.66	2.58	2.49	2.40	2.31	2.21	.010	
3.59	3.42	3.25	3.06	2.97	2.87	2.77	2.66	2.55	2.43	.005	
1.87	1.82	1.77	1.72	1.69	1.66	1.63	1.59	1.56	1.52	.100	25
2.24	2.16	2.09	2.01	1.96	1.92	1.87	1.82	1.77	1.71	.050	
2.61	2.51	2.41	2.30	2.24	2.18	2.12	2.05	1.98	1.91	.025	
3.13	2.99	2.85	2.70	2.62	2.54	2.45	2.36	2.27	2.17	.010	
3.54	3.37	3.20	3.01	2.92	2.82	2.72	2.61	2.50	2.38	.005	
1.86	1.81	1.76	1.71	1.68	1.65	1.61	1.58	1.54	1.50	.100	26
2.22	2.15	2.07	1.99	1.95	1.90	1.85	1.80	1.75	1.69	.050	
2.59	2.49	2.39	2.28	2.22	2.16	2.09	2.03	1.95	1.88	.025	
3.09	2.96	2.81	2.66	2.58	2.50	2.42	2.33	2.23	2.13	.010	
3.49	3.33	3.15	2.97	2.87	2.77	2.67	2.56	2.45	2.33	.005	
1.85	1.80	1.75	1.70	1.67	1.64	1.60	1.57	1.53	1.49	.100	27
2.20	2.13	2.06	1.97	1.93	1.88	1.84	1.79	1.73	1.67	.050	
2.57	2.47	2.36	2.25	2.19	2.13	2.07	2.00	1.93	1.85	.025	
3.06	2.93	2.78	2.63	2.55	2.47	2.38	2.29	2.20	2.10	.010	
3.45	3.28	3.11	2.93	2.83	2.73	2.63	2.52	2.41	2.29	.005	
1.84	1.79	1.74	1.69	1.66	1.63	1.59	1.56	1.52	1.48	.100	28
2.19	2.12	2.04	1.96	1.91	1.87	1.82	1.77	1.71	1.65	.050	
2.55	2.45	2.34	2.23	2.17	2.11	2.05	1.98	1.91	1.83	.025	
3.03	2.90	2.75	2.60	2.52	2.44	2.35	2.26	2.17	2.06	.010	
3.41	3.25	3.07	2.89	2.79	2.69	2.59	2.48	2.37	2.25	.005	
1.83	1.78	1.73	1.68	1.65	1.62	1.58	1.55	1.51	1.47	.100	29
2.18	2.10	2.03	1.94	1.90	1.85	1.81	1.75	1.70	1.64	.050	
2.53	2.43	2.32	2.21	2.15	2.09	2.03	1.96	1.89	1.81	.025	
3.00	2.87	2.73	2.57	2.49	2.41	2.33	2.23	2.14	2.03	.010	
3.38	3.21	3.04	2.86	2.76	2.66	2.56	2.45	2.33	2.21	.005	
1.82	1.77	1.72	1.67	1.64	1.61	1.57	1.54	1.50	1.46	.100	30
2.16	2.09	2.01	1.93	1.89	1.84	1.79	1.74	1.68	1.62	.050	
2.51	2.41	2.31	2.20	2.14	2.07	2.01	1.94	1.87	1.79	.025	
2.98	2.84	2.70	2.55	2.47	2.39	2.30	2.21	2.11	2.01	.010	
3.34	3.18	3.01	2.82	2.73	2.63	2.52	2.42	2.30	2.18	.005	

Table IV (*Continued*)

De-nomi-nator *df*	*p*	Numerator *df*								
		1	2	3	4	5	6	7	8	9
40	.100	2.84	2.44	2.23	2.09	2.00	1.93	1.87	1.83	1.79
	.050	4.08	3.23	2.84	2.61	2.45	2.34	2.25	2.18	2.12
	.025	5.42	4.05	3.46	3.13	2.90	2.74	2.62	2.53	2.45
	.010	7.31	5.18	4.31	3.83	3.51	3.29	3.12	2.99	2.89
	.005	8.83	6.07	4.98	4.37	3.99	3.71	3.51	3.35	3.22
60	.100	2.79	2.39	2.18	2.04	1.95	1.87	1.82	1.77	1.74
	.050	4.00	3.15	2.76	2.53	2.37	2.25	2.17	2.10	2.04
	.025	5.29	3.93	3.34	3.01	2.79	2.63	2.51	2.41	2.33
	.010	7.08	4.98	4.13	3.65	3.34	3.12	2.95	2.82	2.72
	.005	8.49	5.79	4.73	4.14	3.76	3.49	3.29	3.13	3.01
120	.100	2.75	2.35	2.13	1.99	1.90	1.82	1.77	1.72	1.68
	.050	3.92	3.07	2.68	2.45	2.29	2.17	2.09	2.02	1.96
	.025	5.15	3.80	3.23	2.89	2.67	2.52	2.39	2.30	2.22
	.010	6.85	4.79	3.95	3.48	3.17	2.96	2.79	2.66	2.56
	.005	8.18	5.54	4.50	3.92	3.55	3.28	3.09	2.93	2.81
∞	.100	2.71	2.30	2.08	1.94	1.85	1.77	1.72	1.67	1.63
	.050	3.84	3.00	2.60	2.37	2.21	2.10	2.01	1.94	1.88
	.025	5.02	3.69	3.12	2.79	2.57	2.41	2.29	2.19	2.11
	.010	6.63	4.61	3.78	3.32	3.02	2.80	2.64	2.51	2.41
	.005	7.88	5.30	4.28	3.72	3.35	3.09	2.90	2.74	2.62

SOURCE: A portion of "Tables of percentage points of the inverted (*F*) distribution," *Biometrika*, vol. 33 (1943) by M. Merrington and C. M. Thompson and from Table 18 of *Biometrika Tables for Statisticians*, vol. 1, Cambridge University Press, 1954, edited by E. S. Pearson and H. O. Hartley. Reproduced with permission of the authors, editors, and *Biometrika* trustees.

Table IV (*Continued*)

Numerator *df*

10	12	15	20	24	30	40	60	120	∞	*p*	*df*
1.76	1.71	1.66	1.61	1.57	1.54	1.51	1.47	1.42	1.38	.100	40
2.08	2.00	1.92	1.84	1.79	1.74	1.69	1.64	1.58	1.51	.050	
2.39	2.29	2.18	2.07	2.01	1.94	1.88	1.80	1.72	1.64	.025	
2.80	2.66	2.52	2.37	2.29	2.20	2.11	2.02	1.92	1.80	.010	
3.12	2.95	2.78	2.60	2.50	2.40	2.30	2.18	2.06	1.93	.005	
1.71	1.66	1.60	1.54	1.51	1.48	1.44	1.40	1.35	1.29	.100	60
1.99	1.92	1.84	1.75	1.70	1.65	1.59	1.53	1.47	1.39	.050	
2.27	2.17	2.06	1.94	1.88	1.82	1.74	1.67	1.58	1.48	.025	
2.63	2.50	2.35	2.20	2.03	2.03	1.94	1.84	1.73	1.60	.010	
2.90	2.74	2.57	2.39	2.29	2.19	2.08	1.96	1.83	1.69	.005	
1.65	1.60	1.55	1.48	1.45	1.41	1.37	1.32	1.26	1.19	.100	120
1.91	1.83	1.75	1.66	1.61	1.55	1.50	1.43	1.35	1.25	.050	
2.16	2.05	1.94	1.82	1.76	1.69	1.61	1.53	1.43	1.31	.025	
2.47	2.34	2.19	2.03	1.95	1.86	1.76	1.66	1.53	1.38	.010	
2.71	2.54	2.37	2.19	2.09	1.98	1.87	1.75	1.61	1.43	.005	
1.60	1.55	1.49	1.42	1.38	1.34	1.30	1.24	1.17	1.00	.100	∞
1.83	1.75	1.67	1.57	1.52	1.46	1.39	1.32	1.22	1.00	.050	
2.05	1.94	1.83	1.71	1.64	1.57	1.48	1.39	1.27	1.00	.025	
2.32	2.18	2.04	1.88	1.79	1.70	1.59	1.47	1.32	1.00	.010	
2.52	2.36	2.19	2.00	1.90	1.79	1.67	1.53	1.36	1.00	.005	

Table V Gaussian Deviates

Entries are Values of Z where $Z \sim NID(0,1)$*

1.329−	.238−	.838−	.988−	.445−	.964	.266−	.322−	1.726−	2.252
1.284	.229−	1.058	.090	.050	.523	.016	.277	1.639	.554
.619	.628	.005	.973	.058−	.150	.635−	.917−	.313	1.203−
.699	.269−	.722	.994−	.807−	1.203−	1.163	1.244	1.306	1.210−
.101	.202	.150−	.731	.420	.116	.496−	.037−	2.466−	.794
1.381−	.301	.522	.233	.791	1.017−	.182−	.926	1.096−	1.001
.574−	1.366	1.843−	.746	.890	.824	1.249−	.806−	.240−	.217
.096	.210	1.091	.990	.900	.837−	1.097−	1.238−	.030	.311−
1.389	.236−	.094	3.282	.295	.416−	.313	.720	.007	.354
1.249	.706	1.453	.366	2.654−	1.400−	.212	.307	1.145−	.639
.756	.397−	1.772−	.257−	1.120	1.188	.527−	.709	.479	.317
.860−	.412	.327−	.178	.524	.672−	.831−	.758	.131	.771
.778−	.979−	.236	1.033−	1.497	.661−	.906	1.169	1.582−	1.303
.037	.062	.426	1.220	.471	.784	.719−	.465	1.559	1.326−
2.619	.440−	.477	1.063	.320	1.406	.701−	.128−	.518	.676−
.420−	.287−	.050−	.481−	1.521	1.367−	.609	.292	.048	.592
1.048	.220	1.121	1.789−	1.211−	.871−	.740−	.513	.558−	.395−
1.000	.638−	1.261	.510	.150−	.034	.054	.055−	.639	.825−
.170	1.131−	.985−	.102	.939−	1.457−	1.766	1.087	1.275−	2.362
.389	.435−	.171	.891	1.158	1.041	1.048	.324−	.404−	1.060
.305−	.838	2.019−	.540−	.905	1.195	1.190−	.106	.571	.298
.321−	.039−	1.799	1.032−	2.225−	.148−	.758	.862−	.158	.726−
1.900	1.572	.244−	1.721−	1.130	.495	.484−	.014	.778−	1.483−
.778−	.288−	.224−	1.324−	.072−	.890	.410−	.752	.376	.224−
.617	1.718−	.183−	.100−	1.719	.696	1.339−	.614−	1.071	.386−
1.430−	.953−	.770	.007−	1.872−	1.075	.913−	1.168−	1.775	.238
.267	.048−	.972	.734	1.408−	1.955−	.848−	2.002	.232	1.273−
.978	.520−	.368−	1.690	1.479−	.985	1.475	.098−	1.633−	2.399
1.235−	1.168−	.325	1.421	2.652	.486−	1.253−	.270	1.103−	.118
.258−	.638	2.309	.741	.161−	.679−	.336	1.973	.370	2.277−
.243	.629	1.516−	.157−	.693	1.710	.800	.265−	1.218	.655
.292−	1.455−	1.451−	1.492	.713−	.821	.031−	.780−	1.330	.977
.505−	.389	.544	.042−	1.615	1.440−	.989−	.580−	.156	.052
.397	.287−	1.712	.289	.904−	.259	.600−	1.635−	.009−	.799−
.605−	.470−	.007	.721	1.117−	.635	.592	1.362−	1.441−	.672
1.360	.182	1.476−	.599−	.875−	.292	.700−	.058	.340−	.639−
.480	.699−	1.615	.225−	1.014	1.370−	1.097−	.294	.309	1.389−
.027−	.487−	1.000−	.015−	.119	1.990−	.687−	1.964−	.366−	1.759
1.482−	.815−	.121−	1.884	.185−	.601	.793	.430	1.181−	.426
1.256−	.567−	.994−	1.011	1.071−	.623−	.420−	.309−	1.362	.863
1.132−	2.039	1.934	.222−	.386	1.100	.284	1.597	1.718−	.560−
.780−	.239−	.497−	.434−	.284−	.241−	.333−	1.348	.478−	.169−
.859−	.215−	.241	1.471	.389	.952−	.245	.781	1.093	.240−
.447	1.479	.067	.426	.370−	.675−	.972−	.225−	.815	.389
.269	.735	.066−	.271−	1.439−	1.036	.306−	1.439−	.122−	.336−
.097	1.883−	.218−	.202	.357−	.019	1.631	1.400	.223	.793−
.686−	1.596	.286−	.722	.655	.275−	1.245	1.504−	.066	1.280−
.957	.057	1.153−	.701	.280−	1.747	.745−	1.338	1.421−	.386
.976−	1.789−	.696−	1.799−	.354−	.071	2.355	.135	.598−	1.883
.274	.226	.909−	.572−	.181	1.115	.406	.453	1.218−	.115−

*This table is reproduced with permission from tables of the RAND Corporation.

Table V (*Continued*)

1.752 −	.329 −	1.256 −	.318	1.531	.349	.958 −	.059 −	.415	1.084 −
.291 −	.085	1.701	1.087 −	.443 −	.292 −	.248	.539 −	1.382 −	.318
.933 −	.130	.634	.899	1.409	.883 −	.095 −	.229	.129	.367
.450 −	.244 −	.072	1.028	1.730	.056 −	1.488 −	.078 −	2.361 −	.992 −
.512	.882 −	.490	1.304 −	.266 −	.757	.361 −	.194	1.078 −	.529
.702 −	.472	.429	.664 −	.592 −	1.443	1.515 −	1.209 −	1.043 −	.278
.284	.039	.518 −	1.351	1.473	.889	.300	.339	.206 −	1.392
.509 −	1.420	.782 −	.429 −	1.266 −	.627	1.165 −	.819	.261 −	.409
1.776 −	1.033 −	1.977	.014	.702	.435 −	.816 −	1.131	.656	.061
.044 −	1.807	.342	2.510 −	1.071	1.220 −	.060 −	.764 −	.079	.964 −
.263	.578 −	1.612	.148 −	.383 −	1.007 −	.414 −	.638	.186 −	.507
.986	.439	.192 −	.132 −	.167	.883	.400 −	1.440 −	.385 −	1.414 −
.441 −	.852 −	1.446 −	.605 −	.348 −	1.018	.963	.004 −	2.504	.847 −
.866 −	.489	.097	.379	.192	.842 −	.065	1.420	.426	1.191 −
1.215 −	.675	1.621	.394	1.447 −	2.199	.321 −	.540 −	.037 −	.185
.475 −	1.210 −	.183	.526	.495	1.297	1.613 −	1.241	1.016 −	.090 −
1.200	.131	2.502	.344	1.060 −	.909 −	1.695 −	.666 −	.838 −	.866 −
.498 −	1.202 −	.057 −	1.354 −	1.441 −	1.590 −	.987	.441	.637	1.116 −
.743 −	.894	.028 −	1.119	.598 −	.279	2.241	.830	.267	.156 −
.779	.780 −	.954 −	.705	.361 −	.734 −	1.365	1.297	.142 −	1.387 −
.206 −	.195 −	1.017	1.167 −	.079 −	.452 −	.058	1.068 −	.394 −	.406 −
.092 −	.927 −	.439 −	.256	.503	.338	1.511	.465 −	.118 −	.454 −
1.222 −	1.582 −	1.786	.517 −	1.080 −	.409 −	.474 −	1.890 −	.247	.575
.068	.075	1.383 −	.084 −	.159	1.276	1.141	.186	.973 −	.266 −
.183	1.600	.335 −	1.553	.889	.896	.035 −	.461	.486	1.246
.811 −	2.904 −	.618	.588	.533	.803	.696 −	.690	.820	.557
1.010 −	1.149	1.033	.336	1.306	.835	1.523	.296	.426 −	.004
1.453	1.210	.043 −	.220	.256 −	1.161 −	2.030 −	.046 −	.243	1.082
.759	.838 −	.877 −	.177 −	1.183	.218 −	3.154 −	.963 −	.822 −	1.114 −
.287 −	.278	.454 −	.897	.122 −	.013	.346	.921	.238	.586 −
.669 −	.035	2.077 −	1.077	.525	.154 −	1.036 −	.015	.220 −	.882
.392	.106	1.430 −	.204 −	.326 −	.825	.432 −	.094 −	1.566 −	.679 −
.337 −	.199	.160 −	.625	.891 −	1.464 −	.318 −	1.297	.932	.032 −
.369	1.990 −	1.190 −	.666	1.614 −	.082	.922	.139 −	.833 −	.091
1.694 −	.710	.655 −	.546 −	1.654	.134	.466	.033	.039 −	.838
.985	.340	.276	.911	.170 −	.551 −	1.000	.838 −	.275	.304 −
1.063 −	.594 −	1.526 −	.787 −	.873	.405 −	1.324 −	.162	.163 −	2.716 −
.033	1.527 −	1.422	.308	.845	.151 −	.741	.064	1.212	.823
.597	.362	3.760 −	1.159	.874	.794 −	.915 −	1.215	1.627	1.248 −
1.601 −	.570 −	.133	.660 −	1.485	.682	.898 −	.686	.658	.346
.266 −	1.309 −	.597	.989	.934	1.079	.656 −	.999 −	.036 −	.537 −
.901	1.531	.889 −	1.019 −	.084	1.531	.144 −	1.920 −	.678	.402 −
1.433 −	1.008 −	.990 −	.090	.940	.207	.745 −	.638	1.469	1.214
1.327	.763	1.724 −	.709 −	1.100 −	1.346 −	.946 −	.157 −	.522	1.264 −
.248 −	.788	.577 −	.122	.536 −	.293	1.207	2.243 −	1.642	1.353
.401 −	.679 −	.921	.476	1.121	.864 −	.128	.551 −	.872 −	1.511
.344	.324 −	.686	1.487 −	.126 −	.803	.961 −	.183	.358 −	.184 −
.441	.372 −	1.336 −	.062	1.506	.315 −	.112 −	.452 −	1.594	.264 −
.824	.040	1.734 −	.251	.054	.379 −	1.298	.126 −	.104	.529 −
1.385	1.320	.509 −	.381 −	1.671 −	.524 −	.805 −	1.348	.676	.799

Table V (*Continued*)

1.556	.119	.078−	.164	.455−	.077	.043−	.299−	.249	.182−
.647	1.029	1.186	.887	1.204	.657−	.644	.410−	.652−	.165−
.329	.407	1.169	2.072−	1.661	.891	.233	1.628−	.762−	.717−
1.188−	1.171	1.170−	.291−	.863	.045−	.205−	.574	.926−	1.407
.917	.616−	1.589−	1.184	.266	.559	1.833−	.572−	.648−	1.090−
.414	.469	.182−	.397	1.649	1.198	.067	1.526−	.081−	.192−
.107	.187−	1.343	.472	.112−	1.182	.548	2.748	.249	.154
.497−	1.907	.191	.136	.475−	.458	.183	1.640−	.058−	1.278
.501	.083	.321−	1.133	1.126	.299−	1.299	1.617	1.581	2.455−
1.382−	.738−	1.225	1.564	.363−	.548−	1.070	.390	1.398−	.524
.590−	.699	.162−	.011−	1.049	.689−	1.225	.339	.539−	.445−
1.125−	1.111	1.065−	.534	.102	.425	1.026−	.695	.057−	.795
.849	.169	.351−	.584	2.177	.009	.696−	.426−	.692−	1.638−
1.233−	.585−	.306	.773	1.304	1.304−	.282	1.705−	.187	.880−
.104	.468−	.185	.498	.624−	.322−	.875−	1.478	.691−	.281−
.261	1.883	.181−	1.675	.324−	1.029−	.185−	.004	.101−	1.187−
.007−	1.280	.568	1.270−	1.405	1.731	2.072	1.686	.728	.417−
.794	.111−	.040	.536−	.976−	2.192	1.609	.190−	.279−	1.611−
.431	2.300−	1.081−	1.370−	2.943	.653	2.523−	.756	.886	.983−
.149	1.294	.580−	.482	1.449−	1.067−	1.996	.274−	.721	.490
.216−	1.647−	1.043	.481	.011−	.587−	.916−	1.016−	1.040−	1.117−
1.604	.851−	.317−	.686−	.008−	1.939	.078	.465−	.533	.652
.212−	.005	.535	.837	.362	1.103	.219	.488	1.332	.200−
.007	.076−	1.484	.455	.207−	.554−	1.120	.913	.681−	1.751
.217−	.937	.860	.323	1.321	.492−	1.386−	.003−	.230−	.539
.649−	.300	.698−	.900	.569	.842	.804	1.025	.603	1.546−
1.541−	.193	2.047	.552−	1.190	.087−	2.062	2.173−	.791−	.520−
.274	.530−	.112	.385	.656	.436	.882	.312	2.265−	.218−
.876	1.498−	.128−	.387−	1.259−	.856−	.353−	.714	.863	1.169
.859−	1.083−	1.288	.078−	.081−	.210	.572	1.194	1.118−	1.543−
.015−	.567−	.113	2.127	.719−	3.256	.721−	.663−	.779−	.930−
1.529−	.231−	1.223	.300	.995−	.651−	.505	.138	.064−	1.341
.278	.058−	2.740−	.296−	1.180−	.574	1.452	.846	.243−	1.208−
1.428	.322	2.302	.852−	.782	1.322−	.092−	.546−	.560	1.430−
.770	1.874−	.347	.994	.485−	1.179−	.048	1.324−	1.061	.449
.303−	.629−	.764	.013	1.192−	.475−	1.085−	.880−	1.738	1.225−
.263−	2.105−	.509	.645−	1.362	.504	.755−	1.274	1.448	.604
.997	1.187−	.242−	.121	2.510	1.935−	.350	.073	.458	.446−
.063−	.475−	1.802−	.476−	.193	1.199−	.339	.364	.684−	1.353
.168−	1.904	.485−	.032−	.554−	.056	.710−	.778−	.722	.024−
.366	.491−	.301	.008−	.894−	.945−	.384	1.748−	1.118−	.394
.436	.464−	.539	.942	.458−	.445	1.883−	1.228	1.113	.218−
.597	1.471−	.434−	.705	.788−	.575	.086	.504	1.445	.513−
.805−	.624−	1.344	.649	1.124−	.680	.986−	1.845	1.152−	.393−
1.681	1.910−	.440	.067	1.502−	.755−	.989−	.054−	2.320−	.474
.007−	.459−	1.940	.220	1.259−	1.729−	.137	.520−	.412−	2.847
.209	.633−	.299	.174	1.975	.271−	.119	.199−	.007	2.315−
1.254	1.672	1.186−	1.310−	.474	.878	.725−	.191−	.642	1.212−
1.016−	.697−	.017	.263−	.047−	1.294−	.339−	2.257	.078−	.049−
1.169−	.355−	1.086	.199−	.031	.396	.143−	1.572	.276	.027

Table V (*Continued*)

.856−	.063−	.787	2.052−	1.192−	.831−	1.623	1.135	.759	.189−
.276−	1.110−	.752	1.378−	.583−	.360	.365	1.587	.621	1.344
.379	.440−	.858	1.453	1.356−	.503	1.134−	1.950	1.816−	.283−
1.468	.131	.047	.355	.162	1.491−	.739−	1.182−	.533−	.497−
1.805−	.772−	1.286	.636−	1.312−	1.045−	1.559	.871−	.102−	.123−
2.285	.554	.418	.577−	1.489−	1.255−	.092	.597−	1.051−	.980−
.602−	.399	1.121	1.026−	.087	1.018	1.437−	.661	.091	.637−
.229	.584−	.705	.124	.341	1.320	.824−	1.541−	.163−	2.329
1.382	1.454−	1.537	1.299−	.363	.356−	.025−	.294	2.194	.395−
.978	.109	1.434	1.094−	.265−	.857−	1.421−	1.773−	.570	.053−
.678−	2.335−	1.202	1.697−	.547	.201−	.373−	1.363−	.081−	.958
.366−	1.084−	.626−	.798	1.706	1.160−	.838−	1.462	.636	.570
1.074−	1.379−	.086	.331−	.288−	.309−	1.527−	.408−	.183	.856
.600−	.096−	.696	.446	1.417	2.140−	.599	.157−	1.485	1.387
.918	1.163	1.445−	.759	.878	1.781−	.056−	2.141−	.234−	.975
.791−	.528−	.946	1.673	.680−	.784−	1.494	.086−	1.071−	1.196−
.598	.352−	.719	.341−	.056	1.041−	1.429	.235	.314	1.693−
.567	1.156−	.125−	.534−	.711	.511−	.187	.644−	1.090−	1.281−
.963	.052	.037	.637	1.335−	.055	.010	.860−	.621−	.713
.489	.209−	1.659	.054	1.635	.169	.794	1.550−	1.845	.388−
1.627−	.017−	.699	.661	.073−	.188	1.183	1.054−	1.615−	.765−
1.096−	1.215	.320	.738	1.865	1.169−	.667−	.674−	.062−	1.378
2.532−	1.031	.799−	1.665	2.756−	.151−	.704−	.602	.672−	1.264
.024	1.183−	.927−	.629−	.204	.825−	.496	2.543	.262	.785−
.192	.125	.373	.931−	.079−	.186	.306−	.621	.292−	1.131
1.324−	1.229−	.648−	.430−	.811	.868	.787	1.845	.374−	.651−
.726−	.746−	1.572	1.420−	1.509	.361−	.310−	3.117−	1.637	.642
1.618−	1.082	.319−	.300	1.524	.418−	1.712−	.358	1.032−	.537
1.695	.843	2.049	.388	.297−	1.077	.462−	.655	.940	.354−
.790	.605	3.077−	1.009	.906−	1.004−	.693	1.098−	1.300	.549
1.792	.895−	.136−	1.765−	1.077	.418	.150−	.808	.697	.435
.771	.741−	.492−	.770−	.458−	.021−	1.385	1.225−	.066−	1.471−
1.438−	.423	1.211−	.723	.731−	.883	2.109−	2.455−	.210−	1.644
.294−	1.266	1.994−	.730−	.545	.397	1.069	.383−	.097−	.985−
1.966−	.909	.400	.685	.800−	1.759	.268	1.387	.414−	1.615
.999−	1.587	1.423	.937	.943−	.090	1.185	1.204−	.300	1.354−
.581	.481	2.400−	.000	.231	.079	2.842−	.846−	.508−	.516−
.370	1.452−	.580−	1.462−	.972−	1.116	.994−	.374	3.336−	.058−
.834	1.227−	.709−	1.039−	.014−	.383−	.512−	.347−	.881	.638−
.376−	.813−	.660	1.029−	.137−	.371	.376	.968	1.338	.786−
1.621−	.815	.544−	.376−	.852−	.436	1.562	.815	1.048−	.188
.163	.161−	2.501	.265−	.285−	1.934	1.070	.215	.876−	.073
1.786	.538−	.437−	.324	.105	.421−	.410−	.947−	.700	1.006−
2.140	1.218	.351−	.068−	.254	.448	1.461−	.784	.317	1.013
.064	.410	.368	.419	.982−	1.371	.100	.505−	.856	.890
.789	.131−	1.330	.506	.645−	1.414−	2.426	1.389	.169−	.194−
.011−	.372−	.699−	2.382	1.395−	.467−	1.256	.585−	1.359−	1.804−
.463−	.003	1.470−	1.493	.960	.364	1.267−	.007−	.616	.624
1.210−	.669−	.009	1.284	.617−	.355	.589−	.243−	.015−	.712−
1.157−	.481	.560	1.287	1.129	.126−	.006	1.532	1.328	.980

Table V (*Continued*)

.240	1.774	.210	1.471−	1.167	1.114−	.182	.485−	.318−	1.156
.627	.758−	.930−	1.641	.162	.874−	.235−	.203	.724−	.155−
.594−	.098	.158	.722−	1.385	.985−	1.707−	.175	.449	.654
1.082	.753−	1.944−	1.964−	2.131−	2.796−	1.286−	.807	.122−	.527
.060	.014−	1.577	.814−	.633−	.275	.087−	.517	.474	1.432−
.013−	.402	.086−	.394−	.292	2.862−	1.660−	1.658−	1.610	2.205−
1.586	.833−	1.444	.615−	1.157−	.220−	.517−	1.668−	2.036−	.850−
.405−	1.315−	1.355−	1.331−	1.394	.381−	.729−	.447−	.906−	.622
.329−	1.701	.427	.627	.271−	.971−	1.010−	1.182	.143−	.844
.992	.708	.115−	1.630−	.596	.499	.862−	.508	.474	.974−
.296	.390−	2.047	.363−	.724	.788	.089−	.930	.497−	.058
2.069−	1.422−	.948−	1.742−	1.173−	.215	.661	.842	.984−	.577−
.211−	1.727−	.277−	1.592	.707−	.327	.527−	.912	.571	.525−
.467−	1.848	.263−	.862−	.706	.533−	.626	.200−	2.221−	.368
1.284	.412	1.512	.328	.203	1.231−	1.480−	.400−	.491−	.913
.821	1.503−	1.066−	1.624	1.345	.440	1.416−	.301	.355−	.106
1.056	1.224	.281	.098−	1.868	.395−	.610	1.173−	1.449−	1.171
1.090	.790−	.882	1.687	.009−	2.053−	.030−	.421−	1.253	.081−
.574	.129	1.203	.280	1.438	2.052−	.443−	.522	.468	1.211−
.531−	2.155	.334	.898	1.114−	.243	1.026	.391	.011−	.024−
.896	.181	.941−	.511−	.648	.710−	.181−	1.417−	.585−	.087
.042	.579	.316−	.394	1.133	.305−	.683−	1.318−	.050−	.993
2.328	.243−	.534	.241	.275	.060	.727	1.459−	.174	1.072−
.486	.558−	.426	.728	.360−	.068−	.058	1.471	.051−	.337
.304−	.309−	.646	.309	1.320−	.311	1.407−	.011−	.387	.128
2.319−	.129−	.866	.424−	.236	.419	1.359−	1.088−	.045−	1.096
1.098	.875−	.659	1.086−	.424−	1.462−	.743	.787−	1.472	1.677
.038−	.118−	1.285−	.545−	.140−	1.244	1.104−	.146	.058	1.245
.207−	.746−	1.681	.137	.104	.491−	.935−	.671	.448−	.129−
.333	1.386−	1.840	1.089	.837	1.642−	.273−	.798−	.067	.334
1.190	.547−	1.016−	.540	.993−	.443	.190−	1.019	1.021−	1.276−
1.416−	.749−	.325	.846	2.417	.479−	.655−	1.326−	1.952−	1.234
.622	.661	.028	1.302	.032−	.157−	1.470	.766−	.697	.303−
1.134−	.499	.538	.564	2.392−	1.398−	.010	1.874	1.386	.000
.725	.242−	.281	1.355	.036−	.204	.345−	.395	.753−	1.645
.210−	.611	.219−	.450	.308	.993	.146−	.225	1.496−	.246
.219	.302	.000	.437−	2.127−	.883	.599−	1.516−	.826	1.242
1.098−	.252−	2.480−	.973−	.712	1.430−	.167−	1.237−	.750	.763−
.144	.489	.637−	1.990	.411	.563−	.027	1.278	2.105	1.130−
1.738−	1.295−	.431	.503−	2.327	.007−	1.293−	1.206−	.066−	1.370
.487−	.097−	1.361−	.340−	.204	.938	.148−	1.099−	.252=	.384−
.636−	.626−	1.967	1.677	.331−	.440−	1.440−	1.281	1.070	1.167−
1.464−	1.493−	.945	.180	.672−	.035−	.293−	.905−	.196	1.122−
.561	.375−	.657−	1.304	.833	1.159−	1.501	1.265	.438	.437−
.525−	.017−	1.815	.789	1.908−	.353−	1.383	1.208−	1.135−	1.082
.980	.111−	.804−	1.078−	1.930−	.171	1.318−	2.377	.303−	1.062
.501	.835	.518−	1.034−	1.493−	.712	.421	1.165−	.782	1.484−
1.081	1.176−	.542−	.312	.688	.670	.771−	.090−	.611−	.813−
.148−	1.203−	1.553−	1.244	.826	.077	.128	.772−	1.683	.318
.096	.286−	.362	.888	.551	1.782	.335	2.083	.350	.260

Table VI   Gaussian Deviates

Entries are Values X where $X \sim \text{NID}(0,4)$*

−0.221	−0.540	−0.701	5.511	−2.404	−0.987	−0.158	−0.578	−1.893	0.854
−2.454	−2.816	0.580	−1.068	1.010	1.209	2.234	3.224	3.750	1.285
0.089	0.418	−0.421	2.448	−0.279	1.916	−3.166	−0.773	−0.818	−1.411
0.931	1.345	3.164	0.019	0.767	0.439	−3.412	−0.982	0.520	−0.473
0.361	0.794	0.120	−0.347	2.785	0.980	1.003	−1.796	−1.778	−0.783
−0.559	−2.111	−3.396	4.236	2.764	−1.990	−0.060	−2.488	−0.503	−4.406
−4.816	−1.369	1.856	0.383	0.016	−2.144	−0.187	−1.561	1.441	−2.246
0.784	0.607	0.663	−0.764	−1.395	1.738	−2.055	2.962	−1.616	1.326
2.576	−3.024	0.191	1.084	−3.698	−3.031	−0.517	−0.179	0.681	−0.719
−1.232	1.234	−0.046	1.338	1.726	1.448	2.216	−1.662	2.188	2.308
2.129	−1.936	3.381	1.319	−3.131	−1.037	1.191	1.449	0.690	−0.251
−2.753	1.049	1.616	1.232	2.910	0.389	−3.766	2.044	1.459	−0.002
0.071	1.869	−5.827	0.866	−1.191	2.508	1.552	−1.052	1.914	−0.274
0.507	0.595	−0.202	−0.775	−1.732	−2.771	0.049	5.221	3.059	0.015
0.384	−1.574	1.414	−0.789	−1.263	−0.470	0.020	1.489	0.497	1.316
1.688	4.311	2.305	−5.632	1.776	1.540	0.208	0.611	0.810	−1.241
−0.045	−1.563	3.687	−0.160	−0.101	1.838	1.590	1.222	0.377	2.069
2.516	−1.339	0.956	−1.285	0.301	3.739	−3.320	0.183	0.993	−4.678
0.536	1.965	−0.580	−0.307	1.564	0.163	−2.239	−2.460	−2.003	−1.609
0.775	1.427	−0.626	−1.134	−3.109	1.652	2.331	−0.188	2.137	−1.316
0.964	−3.740	1.995	−1.349	−1.068	−0.172	1.907	5.515	0.863	−1.018
2.597	2.328	−0.722	3.057	1.632	0.655	0.972	1.401	1.840	4.508
−1.343	−2.859	0.903	−0.631	−2.810	3.345	1.997	0.356	1.215	0.501
−1.097	0.798	−1.057	3.880	2.321	−1.677	−3.746	−1.125	−1.090	−1.972
−0.977	0.225	−0.004	−0.513	3.613	1.030	2.349	−1.278	−1.301	−5.159
2.421	−1.732	2.170	0.451	1.013	−0.912	0.615	−0.532	1.453	−2.155
0.921	−0.932	−2.511	−0.164	0.154	−0.004	0.516	2.240	−0.020	4.432
−1.854	−3.192	−3.633	0.067	3.709	0.560	−0.156	0.964	−2.618	−0.718
2.457	−0.566	−1.439	0.194	1.440	−1.568	−2.407	−1.356	0.849	0.801
1.143	0.212	4.088	−0.832	−0.361	0.303	−2.984	1.378	−0.649	−2.399
1.184	1.622	−1.896	0.026	−2.163	−1.683	3.778	3.585	−3.853	2.352
3.842	1.179	−0.987	0.498	2.348	3.263	1.924	−4.421	−0.680	2.129
1.930	0.114	−6.145	0.737	−0.353	2.478	−2.104	0.020	−2.250	−1.096
−0.174	−0.403	−1.539	1.740	−1.293	−1.922	1.228	1.433	−2.659	2.923
3.017	2.409	1.876	4.534	−0.539	−1.534	−0.847	−0.107	0.796	−1.257
2.632	−2.417	0.136	−1.155	4.277	−1.035	−0.968	−0.400	1.393	−0.858
−1.589	−2.199	0.776	0.821	3.237	4.810	1.012	−4.102	1.088	1.958
−1.308	0.561	−0.882	4.041	1.923	−0.717	0.599	1.705	0.241	2.677
−1.881	−1.808	−2.767	0.426	0.234	−4.060	1.036	−1.657	0.471	1.753
2.869	4.397	1.986	−2.123	−0.065	−0.705	−0.968	−2.265	−1.979	−1.114
0.252	−1.324	0.302	2.426	0.710	−1.454	−0.319	2.277	−0.971	−3.217
−0.833	1.653	−2.738	2.856	−0.789	−0.873	−0.809	−1.538	−1.334	2.289
0.276	0.020	−0.162	−1.720	0.048	0.401	−2.073	2.430	2.776	1.174
0.742	3.058	1.994	3.090	0.170	−0.789	−2.526	−0.980	−1.331	−1.834
0.770	1.419	4.391	1.502	−2.856	−3.648	−1.179	1.556	3.176	2.613
−2.208	1.766	−0.282	3.051	−1.734	−0.032	1.234	−0.626	1.052	−3.146
−1.161	−0.803	5.530	2.219	−0.371	1.372	−1.649	−2.059	1.456	0.677
−4.276	−0.196	−1.456	0.139	0.094	2.367	−1.902	1.123	−1.222	0.323
−1.615	−0.140	0.697	−0.647	1.289	1.416	0.811	0.523	1.406	−1.022
−3.831	−0.105	−2.271	−3.207	0.539	−1.010	2.646	−1.985	0.347	0.712

* This table is reproduced with permission from tables of the RAND Corporation.

Table VI (*Continued*)

−0.686	0.678	−0.150	−0.334	5.096	0.708	0.403	1.538	1.217	−2.456
0.106	−0.018	2.558	−0.049	−1.061	−0.574	−0.510	−1.036	−0.168	2.516
−0.638	3.191	−4.587	−0.499	−0.510	−1.521	1.325	−1.355	−1.036	−3.305
−3.088	−0.998	0.883	2.230	0.603	0.960	−2.303	0.152	1.423	2.706
1.017	−2.496	−4.981	1.769	−0.252	3.018	2.764	1.105	−1.527	−1.187
0.132	−1.502	−1.748	−3.513	0.878	−0.483	−0.354	−1.103	−2.178	2.192
1.217	−2.380	−0.017	−0.072	−2.066	0.251	0.035	1.641	−1.298	2.070
0.540	−3.950	1.287	−0.771	−2.946	−1.181	−2.286	−1.561	0.420	−0.746
−0.395	−1.421	−2.163	0.270	−1.257	5.610	0.287	−1.138	−0.979	−2.224
−3.279	−2.813	1.125	−3.982	−0.102	0.576	−0.531	−0.695	−3.385	−1.045
3.510	3.235	2.623	−2.582	1.274	2.440	0.824	1.983	−0.770	2.489
−0.408	1.922	2.351	−0.146	−0.039	−0.648	−3.041	0.522	−1.659	−0.515
−2.434	1.297	2.813	−0.651	3.409	0.108	1.716	−1.801	0.344	0.923
0.541	0.820	−0.254	2.851	−3.027	1.553	2.218	1.758	1.003	−1.148
−0.813	1.562	2.116	−4.375	−2.289	1.593	0.163	−1.991	0.355	1.364
−0.238	−0.450	0.364	0.677	−0.711	−1.661	1.071	0.682	0.347	0.113
−1.214	−5.369	3.300	0.461	3.197	−0.368	−3.190	2.868	−0.943	0.931
3.521	1.655	2.373	1.993	−1.096	1.875	−1.143	−3.658	2.664	0.378
−1.075	−3.475	−2.069	−0.971	−0.909	−1.796	−0.760	−1.794	−1.576	3.863
−1.213	−1.760	0.397	−0.323	2.659	0.666	−4.368	2.704	1.160	−1.377
0.408	2.865	3.666	−0.433	−1.798	−1.434	−3.688	2.261	−1.084	0.722
−1.809	−2.890	−3.537	−2.701	0.656	0.684	0.905	1.953	2.720	−0.263
−1.633	0.283	−3.937	−0.224	−0.549	0.016	−1.265	−1.650	−1.506	0.504
−1.181	2.578	0.568	0.286	1.152	−0.929	−3.335	0.020	1.171	1.366
0.374	1.225	−0.213	−1.951	0.126	−1.551	−0.147	0.605	2.450	−1.514
−1.828	−3.459	2.624	2.605	0.698	−0.984	−2.289	−3.389	1.647	−2.592
−3.073	1.381	6.111	0.458	−0.792	−0.785	−1.254	2.784	−0.248	−0.965
−0.817	−1.048	−0.603	−0.647	−2.140	1.970	1.612	−2.050	1.926	−3.520
−0.778	−2.697	2.431	−2.011	2.810	0.010	1.830	−1.425	1.425	−2.506
1.866	0.259	−1.360	2.165	1.845	−0.326	2.054	−0.825	5.348	−0.384
0.715	−0.981	−0.126	0.263	−0.692	−3.790	−3.119	−0.547	−1.450	−0.791
0.802	−0.906	−0.726	0.071	3.693	−0.409	1.536	−0.907	4.216	−3.096
−0.363	0.030	−2.423	−0.517	−4.567	1.092	−1.194	0.253	−2.816	0.766
2.878	−2.689	0.797	0.820	0.764	0.366	−0.891	−0.122	0.196	1.052
−2.245	1.324	0.101	0.431	−2.152	0.779	−0.708	0.028	1.317	−1.259
−0.286	0.390	1.204	−4.414	−0.164	−3.724	0.207	1.835	0.334	0.660
−1.434	−0.736	−0.040	0.213	0.215	−0.565	0.915	−0.022	0.487	−0.487
0.004	−1.773	−0.480	0.768	−0.837	−0.513	0.828	4.563	1.298	2.837
1.299	−1.915	0.346	1.037	−2.953	−1.968	−1.704	1.639	2.802	−2.965
−1.694	1.993	1.021	−2.152	0.679	−0.763	0.577	2.860	0.329	−2.702
−0.506	0.328	−2.091	−0.238	2.582	−0.429	1.647	−1.048	−1.367	1.054
0.527	−3.033	−0.893	−0.776	−0.383	−0.708	−0.482	−2.686	1.369	1.040
3.815	−1.282	1.877	−1.177	0.000	−0.059	0.754	0.529	−0.522	2.073
1.838	−0.569	3.556	−0.956	−2.106	0.371	1.806	0.449	−0.867	2.664
0.810	3.394	1.224	−4.428	4.645	−0.644	−2.579	−2.198	2.683	1.338
4.048	−0.706	1.326	2.187	0.611	2.962	−2.137	−0.657	−3.539	−1.526
1.619	−1.214	−0.860	−0.625	−2.534	0.519	−2.539	0.737	−0.622	−0.146
1.732	−3.199	2.104	1.752	0.087	−0.333	3.943	0.037	0.135	1.756
−1.580	−3.456	2.934	0.580	1.665	2.331	−1.413	−1.558	2.144	1.876
1.176	−0.195	0.127	0.060	−3.145	−0.001	0.219	−1.803	−5.255	1.524

Table VI (*Continued*)

−0.625	−1.119	0.772	−1.479	0.164	3.051	0.297	1.904	2.864	3.093
1.375	1.994	1.004	−3.128	−1.517	−2.916	2.196	3.544	−1.858	−1.021
−1.835	0.393	−1.426	−0.469	−0.009	−2.366	0.408	−0.669	−0.266	−0.907
0.764	−2.796	−1.932	−1.144	−4.177	−2.150	4.163	1.003	−1.088	−0.346
−0.913	−4.834	2.310	−0.154	−2.007	−1.741	−3.570	1.361	−0.219	−1.424
−4.228	0.846	−0.794	−1.756	2.621	0.128	−1.369	2.090	−4.471	0.440
−2.724	2.694	−0.585	1.094	2.116	−1.176	0.180	1.438	1.260	−1.730
4.168	−0.764	−0.791	−2.517	−2.103	0.901	0.141	1.796	−4.435	1.711
3.286	2.374	1.605	−0.951	−1.308	−0.903	1.562	3.537	3.340	−1.417
0.197	0.212	1.303	−0.289	1.441	−2.913	−0.606	−1.302	1.281	0.147
0.713	−1.532	−4.409	−2.502	−1.488	1.696	−2.390	0.517	−2.406	−0.457
0.925	−2.267	2.010	−1.381	−2.057	0.988	−0.024	−2.096	0.116	1.383
−3.312	1.604	0.955	−0.184	0.074	−0.714	2.059	−2.293	0.899	−0.837
0.320	−2.893	−1.005	1.527	−0.990	1.930	−1.512	1.333	3.188	−1.555
0.619	−1.545	1.543	−0.207	−0.586	2.409	−2.454	−0.738	−0.060	−1.533
0.119	−0.542	−2.461	−2.475	−1.265	−3.598	0.983	−1.702	−1.735	−4.773
1.814	−0.053	−0.063	−2.921	2.076	−0.535	2.585	−3.066	−0.771	1.553
−2.068	0.648	2.066	0.610	−0.681	0.845	1.349	0.515	−1.106	−3.860
−1.881	−2.033	1.704	1.161	0.316	1.623	−3.370	−0.261	−3.559	−0.647
−2.125	0.620	−0.838	2.278	0.230	2.962	1.925	−2.209	−0.676	0.859
2.054	−2.290	2.264	1.598	2.064	−1.129	−1.381	−1.149	−0.488	0.568
−2.516	−2.190	−0.629	2.361	1.734	0.607	0.935	1.275	3.125	−0.224
−0.143	−1.222	1.061	−2.668	−4.419	0.569	0.259	−0.027	1.989	4.602
−0.007	0.017	−0.811	−0.166	0.850	0.565	0.184	−2.887	1.101	0.192
−1.749	0.231	−2.380	−3.177	−1.077	4.460	0.494	1.941	−0.106	0.015
1.300	−0.289	−2.657	−0.160	−0.490	−0.329	1.602	−1.110	4.204	2.552
0.588	−1.072	0.935	−0.164	0.113	1.139	−0.923	−0.953	0.001	−0.033
1.719	−1.183	−1.051	−0.944	0.734	1.965	2.121	2.213	3.826	−2.004
0.726	1.867	0.624	2.066	−2.792	−2.507	−0.816	−0.569	0.002	−1.934
2.369	−0.361	2.216	−1.500	−0.350	−1.063	−3.979	−3.626	−1.326	−0.488
1.790	−0.290	2.601	6.261	−0.622	−0.534	0.477	0.075	0.167	−2.351
1.801	−2.408	0.408	−2.039	0.175	3.839	3.096	−0.001	2.912	−0.560
0.392	1.600	−0.940	−0.160	−0.885	−1.083	−3.503	1.814	−0.563	−2.682
−4.113	−3.018	0.523	−1.915	−0.722	−2.769	0.210	−0.381	−0.724	−2.013
−0.393	−0.828	−0.102	−2.457	1.702	2.257	−2.473	−1.459	−1.385	−3.669
0.688	−0.214	2.741	2.906	−0.778	1.158	0.713	0.815	−0.670	0.144
1.957	1.104	3.540	2.726	−0.028	−0.181	−1.477	−4.434	0.457	0.057
1.823	−1.371	−4.951	3.333	0.248	1.691	2.311	−2.996	1.573	2.319
−0.277	0.346	−1.354	3.170	0.268	0.773	1.242	3.542	0.940	−0.535
−0.484	1.447	−0.512	−1.379	−0.808	1.014	2.103	0.005	−2.122	0.843
3.209	−1.924	−0.833	1.158	3.203	−0.040	−0.880	−2.217	0.007	0.022
1.834	2.064	−0.319	2.672	1.281	4.921	0.819	0.634	−4.961	−0.739
−1.618	0.705	0.220	−0.177	−0.117	−4.699	2.210	0.035	2.403	−0.816
2.553	1.710	−2.844	−4.619	−4.328	0.459	−2.373	−1.069	2.792	1.942
−1.665	1.627	1.072	−0.902	1.336	3.850	−0.804	1.254	−2.493	−1.101
−0.468	2.177	1.703	1.897	1.139	−1.606	−1.139	−1.435	5.162	−0.146
−1.296	−0.646	0.193	0.534	−0.863	−3.178	2.461	−1.275	0.731	2.983
−3.086	−0.115	2.325	0.088	4.652	2.833	−0.054	−0.670	−2.313	−1.956
−1.324	0.102	0.665	0.878	−1.760	−1.038	0.685	−1.034	0.380	−0.463
1.936	2.264	0.379	4.480	−3.841	−3.992	−3.565	2.558	−0.906	−0.432

Table VI (*Continued*)

−0.660	−0.996	−0.264	−1.823	0.818	−0.410	−1.786	2.399	1.986	0.242
1.785	−0.471	0.082	−3.006	−2.286	−0.222	0.388	−0.110	−0.358	−0.333
0.880	0.224	2.561	2.165	2.974	2.516	−4.148	−0.241	−1.318	−0.677
1.021	3.100	−1.783	−2.063	−2.176	1.959	0.248	0.597	1.394	0.612
−2.420	5.579	2.351	1.601	1.045	−2.857	1.400	3.411	−0.239	−2.323
−2.542	−3.145	−2.432	−0.444	−1.276	−3.342	3.479	2.630	−0.405	1.845
−2.437	0.104	1.110	0.008	−2.173	2.294	−0.529	1.723	−1.609	2.119
3.280	−1.213	−3.063	3.637	−0.038	0.217	−0.790	−1.412	−0.055	−0.612
0.502	−0.767	−1.569	0.386	−1.990	−0.062	3.319	−2.448	−0.445	0.059
−1.684	0.290	3.179	0.158	3.562	−1.929	−1.170	1.179	0.110	3.655
0.023	1.652	3.049	−1.410	−1.447	−0.638	−2.483	2.386	−0.331	1.215
−2.741	−2.313	−2.069	−1.305	−0.934	−7.769	2.654	−1.032	1.489	0.671
−0.746	2.099	−3.225	1.533	−1.741	−1.922	0.895	−2.974	−0.828	1.734
−0.934	−4.158	−3.297	−2.859	−4.026	2.722	−1.268	0.991	−1.196	−0.458
−1.574	0.097	2.122	−3.279	−0.820	0.483	2.196	0.642	−1.488	0.374
1.261	−0.663	0.616	−2.801	1.065	4.845	0.418	−0.226	1.897	3.554
2.030	1.692	0.265	0.511	−1.959	0.247	−1.381	−2.625	0.695	−2.248
−4.452	0.900	−1.646	0.573	0.973	−0.350	2.649	4.114	2.497	0.287
−0.075	−2.069	−0.574	0.001	−0.784	−1.235	−3.191	2.128	1.168	−0.742
0.369	0.919	−2.760	1.878	−5.001	−1.670	0.913	−2.853	0.002	1.885
1.360	−2.214	−2.175	0.193	3.298	−0.103	2.226	0.164	−2.429	−0.580
−3.271	2.845	−0.102	−0.822	−3.646	0.361	−3.188	−1.031	1.846	−1.622
−0.908	−3.907	1.407	0.078	1.324	0.276	−2.805	0.604	1.632	2.413
−1.323	2.717	−0.083	−1.645	1.103	−1.539	−0.173	2.429	−0.343	0.011
1.095	−0.871	1.636	2.345	−3.127	0.500	1.250	−0.072	−3.248	2.603
−1.907	−0.869	−4.388	−0.114	1.890	0.218	0.510	−0.768	−2.610	3.635
1.508	−0.333	−2.433	1.237	−1.733	−2.826	−3.761	−1.125	0.720	−0.832
2.937	1.887	−0.430	−5.194	4.716	−2.950	−0.393	−1.111	0.008	−0.186
−4.706	−1.302	−2.011	−0.124	−2.037	0.140	1.392	−1.869	−2.249	−0.075
5.019	−3.900	1.300	−0.034	−1.679	−0.621	0.285	1.197	−0.871	−1.240
−0.910	−0.495	0.074	3.144	−2.631	−3.152	0.192	−1.073	0.646	4.381
1.304	−1.010	−0.739	−1.028	2.886	−1.418	1.314	0.779	−2.139	0.173
0.371	−5.663	−0.017	−1.551	−3.508	1.305	−0.819	−0.199	−0.331	−0.358
−2.039	−3.961	0.679	2.451	−2.802	1.449	−0.964	−1.170	0.891	1.560
−0.911	1.904	0.062	−2.375	−1.548	−0.361	2.692	3.772	2.005	3.718
−1.720	0.871	3.594	0.889	0.162	0.112	−0.053	−2.597	−1.310	−2.234
−4.091	0.430	2.222	−0.141	0.506	2.751	−0.472	−1.141	1.671	−0.920
−1.797	−1.272	1.847	0.039	0.689	−0.080	1.457	−3.856	1.332	−2.898
−0.719	0.829	2.570	1.107	−0.314	−3.750	1.041	−1.657	−0.233	1.417
−1.890	3.240	1.877	2.552	3.389	0.215	1.979	−0.895	−1.996	0.611
−1.952	−1.276	−2.754	−0.049	−2.916	3.820	0.381	1.337	2.211	3.456
0.502	1.812	−0.577	0.551	−0.257	0.883	4.377	−4.180	−2.266	−0.100
1.971	2.333	−0.945	2.618	2.953	−1.997	1.491	−0.082	2.617	0.749
0.561	−1.506	4.127	0.933	−1.930	0.460	−0.008	0.352	−1.274	0.271
−1.409	−0.638	2.757	0.461	1.331	2.030	−0.846	−1.035	−1.580	−0.772
−2.066	0.218	0.070	−3.420	0.089	−0.084	4.944	−4.285	0.200	0.276
−2.734	3.622	−0.300	1.648	1.328	0.479	−0.498	1.997	2.203	2.792
−1.434	1.441	0.258	−1.893	−2.925	−1.753	0.272	0.747	−0.999	−0.155
−0.071	−4.344	−2.763	4.371	1.547	2.588	2.914	0.261	3.381	5.445
4.574	1.751	3.420	−1.383	0.966	−2.731	3.444	1.410	2.740	−2.011

Table VI (*Continued*)

−1.739	0.276	1.761	0.092	0.820	1.772	−3.258	0.707	−0.578	−1.611
−1.776	−1.482	1.399	1.031	−0.546	−0.204	2.591	2.129	1.615	0.919
−1.894	0.388	1.023	−1.493	1.513	1.003	2.547	−2.443	−1.855	2.898
−2.042	1.064	−2.399	−0.333	−2.141	−1.022	−2.976	−0.485	0.073	−0.891
0.287	0.120	−2.013	0.598	0.001	3.454	2.077	−1.966	−4.187	0.452
0.739	−4.324	0.088	1.124	0.610	0.368	0.953	−0.141	−0.441	3.163
0.749	0.597	2.194	−0.771	1.063	0.246	0.465	−3.122	−1.995	3.015
−1.111	−2.558	0.146	−0.590	−3.278	2.649	−1.299	−1.809	3.938	−1.766
−0.347	2.324	0.083	−0.152	−3.563	−1.062	0.901	0.882	0.865	0.581
−0.105	1.781	−0.775	−0.726	−3.211	−1.200	−2.688	1.639	−0.945	−1.022
1.968	2.056	−4.124	−1.126	−2.798	−1.150	−1.632	−3.405	1.182	1.985
1.614	−1.436	−4.649	−1.168	2.549	0.522	−0.616	2.009	−0.465	1.362
−1.671	−0.907	−0.459	2.880	2.640	−0.751	−2.414	−1.195	−2.334	−0.240
−1.328	0.335	−0.049	−1.903	0.225	−0.140	−1.121	0.820	0.282	0.635
−0.623	−0.823	1.655	1.997	−3.841	3.318	1.035	1.056	2.112	2.166
−2.292	−1.662	2.136	−0.223	1.372	−3.612	−0.276	−4.097	−0.419	−0.017
3.146	1.248	0.090	−1.069	−0.022	1.017	−1.157	1.803	−1.585	−1.526
1.553	−1.369	0.044	0.606	−1.734	−1.443	−3.016	−0.977	3.150	0.264
−3.179	3.510	−2.299	0.371	1.071	−1.044	−1.352	1.740	1.936	0.242
1.701	−0.455	2.119	−1.716	−2.857	−0.991	−1.621	2.934	3.487	0.754
−2.088	1.495	2.961	−2.029	−0.072	−0.664	0.992	1.659	0.834	−2.175
−3.757	0.316	0.763	−3.035	0.907	−3.804	−3.403	3.689	0.901	−2.386
−1.177	−1.422	−4.712	0.235	−1.048	−2.627	0.794	−1.473	2.598	−0.364
−1.380	−1.661	−1.714	−1.396	0.477	1.750	−2.458	−5.077	−0.194	1.093
−0.852	0.562	−0.199	0.802	0.494	−0.294	0.205	0.260	−2.616	4.117
2.591	1.323	0.458	4.020	−1.907	−0.065	−2.786	0.137	0.446	4.368
−2.240	2.744	0.551	−3.005	−2.677	4.492	2.928	0.061	−0.216	2.566
−1.488	−0.163	−0.187	2.081	−0.993	1.160	1.301	−2.236	1.586	0.011
0.622	−0.988	−0.956	−0.484	−0.648	−3.467	−3.778	1.181	1.740	0.092
−0.949	−2.527	−1.934	1.318	0.422	3.848	0.050	−1.448	0.278	3.041
−4.952	0.019	1.793	0.881	0.282	0.621	1.202	−0.373	3.665	3.386
−0.751	3.342	0.969	0.821	1.983	−0.533	−1.273	−2.214	−0.774	−1.210
0.618	−0.688	−2.960	−5.252	−0.543	0.104	−0.468	−3.139	0.594	−1.302
2.371	0.160	1.715	0.319	1.387	5.138	3.883	−1.869	−0.899	−1.019
−1.184	0.047	1.453	−0.889	−1.292	0.197	−0.302	−1.497	−1.838	−0.940
−0.287	2.329	2.028	−1.765	1.669	−1.024	1.600	0.454	3.098	2.275
1.764	−2.839	−1.942	0.008	4.001	0.083	−1.631	2.968	−0.146	−2.079
1.149	−1.571	1.296	1.510	−0.599	0.083	−0.688	6.017	0.012	−1.451
2.984	−1.432	−0.960	−2.124	1.353	0.934	0.666	3.096	2.905	−1.472
−1.701	−0.004	2.710	0.573	2.424	−0.119	−1.410	3.413	−3.588	0.047
−2.333	0.912	−0.773	−2.016	2.253	2.784	3.764	0.559	4.791	1.288
−0.214	2.787	0.095	−3.174	1.460	0.411	0.922	−0.474	3.113	−1.067
1.214	0.785	−2.686	1.909	−1.747	−4.551	0.589	−0.573	−1.364	−2.583
0.878	0.097	1.650	1.437	−1.643	−2.608	1.122	0.538	0.664	−0.323
−0.105	−0.297	3.821	2.105	2.021	−1.922	1.472	0.042	1.403	1.465
−0.593	0.136	0.910	−0.549	−1.472	3.214	−2.273	3.458	1.436	0.500
2.198	2.325	−1.229	−0.276	1.560	−0.482	−0.482	0.455	−0.181	1.417
1.160	0.139	0.997	−0.082	−0.689	0.995	−5.301	0.998	3.413	−1.797
3.024	−1.561	0.982	−1.244	1.407	−0.063	−1.176	2.355	2.006	−4.833
0.955	0.174	−0.401	2.472	0.584	3.811	1.115	0.951	−2.136	−2.324

BIBLIOGRAPHY
AND REFERENCES

ACTON, F. S., *Analysis of Straight-Line Data*. New York: John Wiley & Sons, Inc., 1959.

ALDER, H. L., AND E. B. ROESSLER, *Probability and Statistics* (4th ed.). San Francisco: W. H. Freeman and Company, 1968.

BROWNLEE, K. A., *Statistical Theory and Methodology in Science and Engineering* (2d ed.). New York: John Wiley & Sons, Inc., 1965.

COCHRAN, W. G., AND G. M. COX, *Experimental Designs* (2d ed.). New York: John Wiley & Sons, Inc., 1962.

DAVIES, O. L., *Design and Analysis of Industrial Experiments*. New York: Hafner Publishing Company, 1954.

DIXON, W. J., and F. J. MASSEY, *Introduction to Statistical Analysis* (3d ed.). New York: McGraw-Hill, Inc., 1969.

EHRENFELD, S., AND S. B. LITTAUER, *Introduction to Statistical Methods*. New York: McGraw-Hill, Inc., 1964.

FEDERER, W., *Experimental Design, Theory & Application*. New York: The Macmillan Company, 1955.

FREUND, J., *Modern Elementary Statistics* (3d ed.). Englewood Cliffs, N.J.: Prentice-Hall, Inc., 1967.

————, P. LIVERMORE, AND I. MILLER, *Manual of Experimental Statistics*. Englewood Cliffs, N.J.: Prentice-Hall, Inc., 1960.

FRYER, H., *Elements of Statistics*. New York: John Wiley & Sons, Inc. 1954.

GOULDEN, C. H., *Methods of Statistical Analysis* (2d ed.). New York: John Wiley & Sons, Inc., 1952.

GRAYBILL, F. A., *An Introduction to Linear Statistical Models*, Vol. 1. New York: McGraw-Hill, Inc., 1961.

GRIFFIN, J. I., *Statistics: Methods and Applications*. New York: Holt, Rinehart and Winston, Inc., 1962.

GUENTHER, W. C., *Analysis of Variance*. Englewood Cliffs, N.J.: Prentice-Hall, Inc., 1964.

187

HICKS, C. R., *Fundamental Concepts in the Design of Experiments.* New York: Holt, Rinehart and Winston, 1964.

JOHNSON, N. L., AND LEONE, F., *Statistics and Experimental Design*, Vols. 1 and 2. New York: John Wiley & Sons, Inc., 1964.

KEEPING, E. S., *Introduction to Statistical Inference.* Princeton, N.J.: D. Van Nostrand Company, Inc., 1962.

KEMPTHORNE, O., *The Design and Analysis of Experiments.* New York: John Wiley & Sons, Inc., 1952.

LI, J. C., *Introduction to Experimental Statistics.* New York: McGraw-Hill, Inc., 1964.

MAKSOUDIAN, Y. L., *Probability and Statistics with Applications.* Scranton, Pa.: International Textbook Company, 1969.

MANDEL, J., *The Statistical Analysis of Experimental Data.* New York: Inter-Science Publishers, 1964.

MILLER, R. G. JR., *Simultaneous Statistical Inference.* New York: McGraw-Hill, Inc., 1966.

NEVILLE, A. M., AND J. B. KENNEDY, *Basic Statistical Methods.* Scranton, Pa.: International Textbook Company, 1964.

OSTLE, B., *Statistics in Research.* Ames: Iowa State University Press, 1954.

PENG, K. C., *The Design and Analysis of Scientific Experiments.* Reading, Mass.: Addison-Wesley Publishing Company, Inc., 1966.

SCHEFFÉ, H., *The Analysis of Variance.* New York: John Wiley & Sons, Inc., 1959.

SNEDECOR, G. W., *Statistical Methods* (5th ed.). Ames: Iowa State University Press, 1956.

STEEL, R. G., AND J. H. TORRIE, *Principles and Procedures of Statistics.* New York: McGraw-Hill, Inc., 1960.

WALLIS, W. A., AND H. V. ROBERTS, *The Nature of Statistics.* Toronto: The Free Press-MacMillan, 1962.

———, *Statistics, A New Approach.* Glencoe, Ill.: The Free Press, 1956.

WILLIAMS, E. J., *Regression Analysis.* New York: John Wiley & Sons, Inc., 1959.

WINE, R. L., *Statistics for Scientists and Engineers.* Englewood Cliffs, N.J.: Prentice-Hall, Inc., 1964.

WINER, B. J., *Statistical Principles in Experimental Design.* New York: McGraw-Hill, Inc., 1962.

WORTHAM, A. W., AND T. E. SMITH, *Practical Statistics in Experimental Design.* Dallas, Texas: Dallas Publishing House, 1959.

INDEX

A

Adjusted scores, 110
  illustrative example, 110
Analysis of variance, 72
AOV, definition, 75
  discussion of, 72
  format, 74, 78
  for hierarchal situations, 139
  for randomized block design, 86
  for 3^2 factorial experiments, 128
    illustrative example, 129
  for 2^2 factorial experiments, 119
    illustrative example, 119
  for 2^3 factorial experiments, 123
    illustrative example, 123
Auxiliary variable, 92
Average absolute deviation, 6

B

Basic partitioning rule of AOV, 140
Best procedure, 31
Bias, in S, 48
  in $\hat{\theta}$, 47, 50

C

Chi Square, distribution, 34, 40
  family, 34, 40

Chi Square (*cont.*)
  table, 163
  variable, 41
Coefficients for contrasts, 119, 124, 127
Combined unbiased estimator, 57
Complement, 12
  rule, 13
Completely randomized experiment, 73
  illustrative example, 73
Composite, -versus-composite, 38
  hypothesis, 29, 36
Concept, of a linear fixed-effects model,
    illustrative example, 77
  of a variance-components model,
    illustrative example, 79
Confidence interval, 26, 43
  for β, 105
  for μ, 34
  for the ratio $\dfrac{\sigma_y^2}{\sigma_x^2}$, 67
  for $\delta = \mu_x - \mu_y$, 61, 70
  for σ^2, 42
  for σ^2 in a regression context, 105
Confounded, 117
Contrasts, coefficients, 119, 124, 127
  definition, 115
  interaction, 114
  orthogonal, 115

Contrasts (*cont.*)
  simple, 115
Corrected cross products, 91
Correction factor, 74
Correlation, simple, 90, 91
Covariable, 92
Covariance, 90, 91, 139, 141
Critical, point, 29
  region, 29
Cross, classification, 72, 84, 147
  products, 91

D

Degrees of freedom, 16
  for Chi Square, 40
  pooled, 57
Design, of experiments, 69
  two-way, 84
Distribution of b, 103

E

Equality, of means, 60, 74
  of variances, 64
Estimate, pooled, 57
  of the regression intercept, 95
  of the regression slope, 95
  of $\tau_k - \tau_l$ in a Latin square experiment, 156
  of a variance component, 80
  weighted for μ, 56
  weighted for σ^2, 57
Expected, length of an interval, 43, 70
  mean squares for a random-effects model, 81
  mean squares for randomized block model, 86
  value, 3
Experimental, design, 69, 73, 104, 132
  error, 69, 109, 118
  units, 2

F

Factor, of an experiment, 115
  one at a time, 67

Factorial, data, 129
  experiments, 115
  3^2 experiments, 125
  2^2 experiments, 117
  2^3 experiments, 122
  2×3 experiments, 137
Field layout, 73
Fixed-effects model, 77, 119, 138
Frequency, curve, 10
  distribution, 10
Fundamental principle, 30

G

Galton, Francis, 95
Gaussian, curve, 13
  deviate tables, 174, 179
Gosset, William, 33

H

Hierarchal classification, 72, 138
Histogram, 9
Homogeneity of experimental units, 67

I

Incomplete, block design, 155
  data, 153
    experiments, illustrative example, 154
Independent random variables, 19, 138
Interaction, 12, 113
  effect, 114
  illustrative example, 114
Interval estimate, 25
  for μ, 34

L

Lack of fit, 109
Latin square experiment, 153, 155
Least squares, 90, 92
  criterion, 94
  regression, illustrative examples, 93, 105

Length of confidence interval for
$\delta = \mu_x - \mu_y$, 61
Linear, combination, 48, 115
contrast, 126
correlation, 90, 91
regression, 90
when may not fit, illustrative example, 109
statistical model, 77

M

Mean, 3
definition of, 49
population, 3, 46
sample, 3
square, 75
square error, 50, 78
Minimum, expected length, 44
variance, 49
variance estimates, 57
Missing observations, 155
Mixed-effects model, 86, 149, 157
Model, fixed-effects, 77, 117
when interaction is present, 115
illustrative example, 115
for linear regression, 97
mixed-effects, 149
random-effects, 79
Mutually exclusive, 12

N

N, fold hierarchy, 139
by N Latin square, 153
-way cross-classification, 147
illustrative example, 147
Nested classification, 72, 138
Normal curve, 13
Null, event, 13
hypothesis, 28

O

One-way experiments with unequal numbers per classification, illustrative example, 80

Orthogonal, contrasts, 115, 121, 123, 126
sources of variability, 102, 157
Overparameterized, 117

P

Paired data experiments, illustrative example, 68
Paired experiments, 67
Parameter, 3
space, 29
Point estimation, 46, 131
Pooled, degrees of freedom, 57
estimator of σ^2, 57
within mean square, 75
Population, 3
mean, 3
variance, 6
Probability, 11
addition rule, 13
of errors in a composite-versus-composite situation, illustrative example, 39
of a type II error, 36

Q

Quadratic contrast, 126

R

Random, -effects model, 79
error in regression, 99
sample, 9
variable, 9
Randomized block design, 84, 147
Range, 6
Reduction in sum of squares due to considering the mean, 72
Regression, curve, 98
linear, 90
model, 99
Restrictions on parameters, 117

S

Sample, correlation coefficient, 91
  covariance, 55, 91
  mean, 3
  random, 9
  size, 24, 70
  variance, 6
Simple, contrast, 115
  hypothesis, 29
Snedecor F, distribution, 63
  family, 63
  relationship to Student t, 65
  table, 166
Standard, deviation, 7
  normal, 12, 15
  normal table, 160
Statistical, hypothesis, 28
  inference, 19
  layout, 1, 73, 80
Student's, family, 33
  relationship to Snedecor F, 65
  t distribution, 33
  t table, 164
  test for equality of mean, 60
Sum of squares associated with a contrast, 118, 126
Summation, notation, 4
  sign, 3
Symmetric distribution, 47

T

Test, of $\beta = 0$, 104
  for equality of means, 60
  for equality of variances, 64
  of hypothesis, 28
  statistic, 35
  statistic for a hypothesis assuming a random model, 143

Test (cont.)
  statistic for an interaction hypothesis, 124
  statistic for paired data, 68
  two-tailed, 37, 60
  of variance, 44
Testing a simple hyp versus simple hyp, illustrative example, 31
Treatment combinations, 116
Two-way cross-classification design, illustrative example, 85
Type, I error, 29
  II error, 29
Typical value, 3

U

Unbiased, 20
  estimator, 47
Unequal numbers per classification, 80
Union, 12

V

Variability of $\bar{X}$, illustrative example, 22
Variance, components, 79, 158
  components model, 79
  of a difference, 54
  minimum, 49
  population, 7
  sample, 6
  of a sum, 54
  of $\bar{X}$, 21, 49

W

Weighted, estimate of μ, 56
  estimate of σ^2, 57